FIRE STRATEGIES— STRATEGIC THINKING

BY PAUL BRYANT

This book is published by Amazon Createspace in conjunction with Kingfell publications, and is the copyright of Kingfell Publishing, London. www.kingfell.com 2013

Copyright © 2013 Paul Bryant
All rights reserved.

ISBN-10: 1482572621
EAN-13: 9781482572629

I would like to thank Dr Nicole Hoffmann, Jose Maria Sanchez de Muniain and Dr Ahmed Allam for their assistance and support while I was writing this book. I would also like to thank my Father, Alan, for some of the military comparison ideas.

I would like to dedicate this book to Kaajal, Ruby and Poppy.

I would also like to dedicate this book to my friends and colleagues who worked with me at Kingfell, past and present – thank you.

Some of the proceeds from the sale of this book will go towards the chosen charities of the UK's Worshipful Company of Firefighters.

www.firefighterscompany.org

ABOUT THIS BOOK

The book has been specifically written for an international audience even though it uses examples from Paul's UK and European experiences. It is not designed to be highly technical but more thought-provoking. The contents are not just applicable to fire safety professionals but anyone who is involved with the fire safety of buildings, including enforcers, architects, structural and building engineers, building control officers, insurers, risk managers, and those responsible for fire safety in organizations.

It can also assist with the thinking process for those in academia and studying for a fire safety qualification.

CONTENTS

Preface..................................xiii

Chapter 1. Think Strategically...............1

Chapter 2. Our Entente Cordiale: Teamwork...13

Chapter 3. The Mission....................33

Chapter 4. The Rules of Engagement........53

Chapter 5. Using Intelligence...............73

Chapter 6. Know the Enemy................83

Chapter 7. The Physical Terrain............99

Chapter 8. The Human Terrain............117

Chapter 9. Strategic Vision................135

Chapter 10. Using Your Resources..........157

Chapter 11. Heading to Victory.............169

FIGURES

Note that all figures are the copyright of Kingfell. They all either appeared in British Standard Specification PAS 911: 2007 for the first time or were published prior to that publication. In all cases, they have been redrawn for the purposes of this book.

Figure 1: The fire strategy objectives matrix 43

Figure 2: Analogy of prescriptive vs. performance-based solutions 62

Figure 3: Fire risk profile curves 96

Figure 4: Significant hazard matrix 114

Figure 5: Significant hazard matrix as used 115

Figure 6: Fire strategy value grid 137

Figure 7: Use of fire strategy value grid 154

Figure 8: Marginal value of fire precautions 160

Figure 9: Quantified assessment of options 164

Figure 10: Components of a fire strategy flow chart . . 166

PREFACE

"Fire safety engineering is a beautiful subject."
—Dr Ahmed Allam, Visiting Professor - Ulster University

I loved the simplicity of the opening quotation. The term "beautiful subject" conjures up passion and artistry. I have also known this subject referred to as a "black art."

Fire engineering, or fire *safety* engineering, encompasses a whole host of scientific and engineering disciplines from mathematics, chemistry, physics, and most branches of engineering through to computer science and the psychology of human behaviour.

It deals with an adversary that is as old as the universe, fundamental in both creation and destruction. This adversary is one that acts just like any living being, hard to predict and potentially dangerous. It is a subject matter worthy of its own branch of study.

Based on the growth of fire engineering degree courses around the world, it is one of the more modern of engineering subjects. Whereas once upon a time, people entered the profession via another science or engineering subject, many students now choose the subject for their first degree.

It is unique as a profession, in that there can be almost as many viewpoints as practitioners when it comes to the best application of fire safety and protection. "Opinion engineering" is an apt way to describe it. There is also a view that the profession is undervalued and that the output of fire engineers is not recognised in the same way as other branches of engineering or science. I will tackle this in more detail within this book.

The subject of fire strategies has become a passion of mine for a large part of my life, a passion that started whilst working in Europe.

Back in the mid-1980s, I was involved in European insurance standards making, as part of my role with an organisation, now defunct, called the "Fire Offices' Committee". The FOC, as it was widely known, was an authority set up by UK Insurers in 1868 to help set tariffs for fire insurance. Over the years the FOC was proactive in setting up approval schemes for fire protection systems and wrote a number of "Rules," including those for sprinkler systems, fire detection and alarm systems, and passive fire protection.

I was representing the United Kingdom Fire Insurers at meetings of the *Comite Europeen des Assurances*. This organization promoted the European interests of insurance companies and, amongst many other tasks, was responsible for the preparation of European technical standards. Meetings of the Fire Group were mostly held at insurance headquarters in the Opera district of Paris. We were a small team made

PREFACE

up from different member countries, and all with a similar background to my own—engineers involved in the approval of fire detection and alarm systems.

Our objective was to draft a fire detection system standard to be used by European insurance companies, primarily to ensure that adequate levels of property protection were in place for "risks" underwritten by the companies. One of our tasks was to draft text to describe the preferred sequence of events that followed the initial detection of a fire, to ensure that the fire would be extinguished either manually or via fire extinguishing systems. We also had to describe how the sequence related to life safety measures. The "core" committee consisted of members from France, Germany, Sweden, Austria, and the United Kingdom, with other countries participating on an "as and when" basis. After a long *dejeuner français*, we were at an impasse. How could we clearly describe the chain of events in a manner that was instantly understandable by all around the table? Furthermore, whatever we did agree on had to be translatable into the three main languages.

I've always had the habit of doodling and, on this day, started to draw a flow chart showing the interaction of fire safety and protection systems, from fire detection right through to two fundamental objectives of any fire strategy: (1) getting people out the building and (2) putting the fire out. This doodle was discussed at the meeting, and with the help of my European colleagues, we translated the drawing into text that we all understood. It was at this point I realised that a "holistic" or "strategic" approach can help focus on the key issues prior to detailed analysis.

FIRE STRATEGIES - STRATEGIC THINKING

The doodle itself was never used, but I did keep it to one side and used it some years later when Kingfell was offered the contract to draft one section of a UK standard covering the application of fire safety engineering principles for buildings (British Standard BS 7974). This same diagram is included within this book.

Those who have ever studied business, whether for a degree or MBA, will recognise that a core topic of this qualification is to understand the processes in developing a *business* strategy. This is a well-documented subject, with much of the research and development deriving from the United States. What is apparent in such studies is the general use of "tools" for use in the evaluation of businesses. These tools are mostly conceptual diagrams or graphs allowing various corporate properties to be assessed or plotted in a clear and concise manner, thus allowing comparative strengths and weaknesses to be highlighted. It became clear to me that such tools could provide equally useful methods in assessing the parameters of fire safety and fire protection, and could lead to a more consistent approach in specifying fire engineered processes and systems.

Moving forward to 1996, and a few months after I formed Kingfell, with little in the way of work, I decided to begin my quest in developing "strategic tools" for fire protection system evaluation. I was fortunate to know many of the key experts in the fire industry via my involvement in British Standards making, and thus had persons of the necessary ability to bounce my ideas off. With the help and support of the then editor of a fire safety journal, John Northey (a guru in the British fire industry but now sadly passed away), these

PREFACE

ideas were published in two monthly instalments of the journal in August and September of that year under the title "Fire protection strategic appraisal tools—parts 1 and 2." The term fire protection was used as the articles concentrated more on system concepts rather than management processes.

At the time, the term "fire strategy" was not as widely used as it is now; thus, there was not the immediate take-up of the idea that I may have initially hoped for. However, I do remember attending a lecture some two years later, and, there in front of me, was a fire consultant using *my* model to explain *his* theories!

Some of the strategic tools I have prepared are included within this book and are taken from my work, predominantly incorporated in British Standard Specification PAS (Publicly Available Specification) 911 published in 2007.

My original reason for writing PAS 911 was to contain my thoughts and ideas within one document. Prior to writing the PAS, a number of concepts appeared in various articles, workshops, and conferences. I approached British Standard Institution in 2006 with my proposal. They did some background research to ensure that the subject matter had not already been covered.

Once it was agreed that the subject of fire strategies was sufficiently different from other codes and standards, I went to work, collating my thoughts and ideas developed over the previous two decades and adding some new ideas along the way. There was one concept, however, that eluded me. I wanted to explore the idea of capturing the essence of a fire strategy simply and coherently, preferably on one side of A4 paper.

FIRE STRATEGIES - STRATEGIC THINKING

I realised that the best way of achieving this was to come up with a simple diagram illustrating all the main elements of a fire strategy and to allow the user of the diagram to identify how each element contributes to the overall fire strategy. It wasn't until January 2007 when the idea hit me, while in Antalya, Turkey, during my forced seven-day exile to enable me to complete the PAS without disturbance.

After supping a number of Turkish coffees, it dawned on me that a simple spider diagram was the answer. The diagram is in this book—Figure 6. PAS 911 was complete and the first full draft was ready by March 2007 and subsequently published in the autumn of that year. I continue to drink Turkish coffee in the hope of another big idea...still waiting!

For the record, the number 911 was *not* chosen to align itself to the New York Tragedy in September 2001. It is the US emergency dialling code. The number "999" had already been reserved for British Standard 9999, published a year or so later. So to the person who wrote to me complaining that the number was "in bad taste"—that is the reason!

Although this book contains some of the ideas of PAS 911, it can be read without the need to refer to the PAS. Its primary aim is to promote the thinking process that, I believe, is most important. The book tries to give simple pointers and highlight where errors can occur, such as where those involved were so busy doing what they love, that they missed the "elephant in the room." It does not purport to be the only method to be used in preparing a fire strategy, but it could be considered as one point of view.

PREFACE

This book will not endlessly refer to specific requirements of regulation and standards, as there is more than sufficient guidance available. The reader may not always agree with what is contained within these pages, but that also is not seen as a bad thing. It is the honest and open exchange of views that will help the fire industry move forward. Fire protection technology is progressing at a fast rate, yet I wonder if the philosophy of fire safety, protection, and fire engineering is keeping up.

CHAPTER 1.
THINK STRATEGICALLY

"However beautiful the strategy, you should occasionally look at the results."
—Winston Churchill.

Every office building, every airport, every shop, every oil and gas processing plant, and, for that matter, every domestic dwelling, is the subject of a fire strategy. The strategy may be implicit or explicit.

For instance, starting with the place where we live, home, the "implicit" fire strategy will be that the building was constructed in accordance with national building regulations (all of which contain fire safety requirements). The electrical system within our home will contain safety systems to prevent overheating or sparks igniting the furnishings. In some cases, a smoke alarm may be fitted to alert sleeping occupants of a fire. All these provisions put together amount to our home's *fire strategy*. Consequently, *everything* constructed for us to live, play, or work in will have a fire strategy.

Now let us take this idea still further, in that all living beings carry around a fire strategy embedded within their inner brain. For reference, the inner brain is one of the oldest parts of the human body, evolutionarily speaking, and it contains basic survival information known as the 4 Fs: fight, flight, food, and fornicate. In most cases, it will be the reaction to run (*flight*), when the person is confronted by fire that could threaten their life. In some cases, the urge will be to stand and *fight* the fire.

And what if any one of us were confronted with the choice of trying to survive within a burning building, or to die some other way? Well, it would appear that the strategy embedded within us would make us choose the latter, as borne out by the examples around the world of "jumpers"—those deliberately choosing to leave burning high-rise buildings by jumping from windows, roofs, or balconies to certain death. Our internal fire strategy is therefore potentially extreme, and one that we will only realise when we are confronted with extreme conditions.

The inference here is that the concept of a fire strategy, when we think about it, is fundamental to all of us, not just those of us who are involved with fire safety.

It is the thinking of what is fundamental in a fire strategy that this book is trying to highlight. We really do need to think first! My experience of the fire safety and protection industry is that we are all so busy doing it, we never get to properly

> **When you think about it, a fire strategy is fundamental to all of us!**

consider *why* we are doing it. It is often a case of act first, review and amend later. A better way is to get it right first time…so let us *think* strategically!

It seems obvious enough that a fire strategy should be strategic. But the strange and unusual thing is that many fire strategies, prepared around the world, are far from strategic.

The term "fire strategy" is widely used, yet it is often misunderstood, even by those operating within the fire safety sector. I remember presenting a one-day course entitled "fire protection strategies" to a small group of fire safety specialists. Before the proceedings, I asked the participants to introduce themselves and state what they wanted from the day. One experienced building control officer explained to all that he heard a lot about fire strategies but had no idea what they were all about. The others in the room nodded with agreement.

If we take the term literally, a fire strategy could indicate that it is a strategy for *creating* fires. Quite obviously, this is not what is meant. In fact, what we really mean is a fire *safety* strategy, or a fire *engineering* strategy, or a fire *safety engineering* strategy. Then there are subsets of the fire strategy, such as fire protection strategy, fire evacuation strategy, and so on. Nevertheless, the two-word term is succinct enough and highlights the main point of the message; it is a strategy to do with fire.

There are many different ways of describing how a set of fire safety and fire protection measures can be encompassed in a specific way, but the term strategy seems to be the most appropriate.

Alternatively, we could use the term tactics, but this term has more to do with the methods required to achieve the strategy.

The term strategy derives from many sources, but notably, it is the military use of the word that is probably most relevant. Definitions or descriptions of what a strategy is include:

- *A long-term plan of action designed to achieve a particular goal*
- *A broad, non-specific statement of an approach to accomplishing desired goals and objectives.*
- *An elaborate and systematic plan of action*
- *A long-term plan for success, to achieve an advantage*

Wars and individual campaigns have often been won or lost by the strategy adopted by the military leaders of the time. In the early nineteenth century, Napoleon Bonaparte, leader of the French armies, was respected around the world as a formidable strategist. He had the ability to comprehend a vast number of factors and put them into a structure that allowed him to beat his enemies, time and time again. He could move massive contingents of soldiers around Europe in a way that their direction and even speed was predetermined to get the maximum force, at the right place and at the right time. The Austrian and Russian armies suffered as a consequence.

Further back in history, Alexander the Great was able to conquer most of the known world in only twelve years due to his ability to decipher information ranging from the abilities of his armies, those of his enemy, and the physical and logistical restraints of the terrain, to develop a clear and concise plan to succeed.

The common theme between these strategists is that they were able to take a great deal of data and formulate a clear and concise plan that would lead to victory.

These days the term strategy is widely used in business, in that all business should prepare a business strategy. A business strategy could be many things, including:

- *Analysis of the market in which the business is operating:* Key considerations here would include the barriers to entry and exit, the stage in the industry lifecycle (e.g. growing, mature, or declining), economic factors, technology factors, external factors posing potential threats, and possible opportunities.
- *Competitive analysis:* Evaluating the type, size, and abilities of the competition; assessing their strengths and weaknesses.
- *Analysis of the business:* Identifying the key attributes of the business itself including strengths, weaknesses, opportunities, and threats.
- *Determining the business plan:* Identifying the main goals for the business in the next year (short term), in five years (medium term), and in ten years (long term); determining the financial requirements to meet this plan; determining the human, material, and technological and infrastructure resource requirements, etc.

The term strategy in business and in war is all about making the best use of knowledge and data to formulate a single plan for success.

FIRE STRATEGIES - STRATEGIC THINKING

Possibly the best way to succinctly encompass what a strategy means is to describe the game of chess. For those persons who play chess, this ancient game of strategy is relatively simple to pick up yet is extremely difficult to master. Each player has an equal and defined number of pieces and plays on a board, which is similarly precise and defined. However, it is the *strategy* of playing that can lead to a great winner. Those who play without any form of strategy are most likely to fail. The strategy of a successful player involves analysing what the opponent is doing and planning a series of moves, in advance, to negate the opponent's attack and at the same time plan a counterattack. As the opponent moves a piece, the player ensures that it can be countered in some way, either immediately or eventually. In the same way, a fire strategy may not be able to eliminate the fire immediately but should ensure that "all bases are covered" so that the fire does not end up as the victor.

In essence, a fire strategy needs to have the following five *properties*:

1. To be *specific* to the unique set of fire-related parameters of the building or structure to which it applies, of the processes within, and of the occupancy profiles. This is later referred to in this book as "the terrain." There is no such thing as a "generic" fire strategy in the same way that there is no such thing as a generic business strategy or a generic military strategy.
2. To be a clear and concise document, despite the necessary and sometimes complex processes

throughout its drafting. It will need to be understood by *all* parties affected by it and not just by fire safety professionals involved in its preparation.
3. To have the necessary level of detail to enable, for instance, fire safety management plans to be drawn up and to provide the design criteria for passive and active fire protection. At the same time, it should not be so detailed that it is inflexible to changing fire safety and protection technologies or philosophies.
4. To have realistic and achievable goals. A strategy will need to take into account practical, logistical, and commercial limitations. A complex "all bells and whistles" strategy may be desirable from the part of a specifier or enforcer, but if it cannot be made a reality, it won't be, and thus the process would mostly be a waste of time and money.
5. A fire strategy is an organic document. It should be modified and adjusted for it to remain true to its inherent goal, and that is preventing and mitigating fire incidents and their impact. Drivers that dictate the need for modification of the document include changes in legislation or stakeholder requirements, revised building structures or layouts, changes in occupancy or use of the building, and new available technology or research. *Past incident reviews show that in too many cases, documents have ended up in somebody's drawer, only to be pulled out after the inevitable has happened. By then they are usually hopelessly out of date and completely defunct.*

FIRE STRATEGIES - STRATEGIC THINKING

An "explicit" fire strategy (i.e. a dedicated document stating the fire safety and protection provisions) is probably most warranted for buildings and structures where simple fire precautions would largely be inadequate. A small office, shop, or residence may require a simple set of provisions. Even if these provisions are based on the outcome of a fire risk assessment, they are likely to be consistent with tried and trusted prescriptive rules and regulations. The fire strategy process, as covered by BS PAS 911, is really designed for more complex building arrangements or for special structures where there is no obvious and quick solution. It will involve a number of steps and could utilise the skills and resources of a number of persons. It will therefore have a cost associated with it. This cost could conceivably be more than the total budget for fire precautions for small premises. Thus, there is an economic consideration, which tends to mean that there is a point where the preparation of an "explicit" fire strategy is just not viable.

Despite this, the real purpose of a fire strategy is to ensure coordination of fire safety provisions, fire safety management, and fire protection systems, which, in theory, should lead to a more effective and achievable solution, optimising the capital spending and the subsequent annual maintenance spending. This is clearly the objective, not just for more complex scenarios, but for all scenarios.

The question of whether or not the preparation of an explicit fire strategy document is justified will need to be decided on a case-by-case basis. Considerations should include:

- *Is the building of "non-standard" design, e.g. does not directly conform to national building regulations?*
- *Is the building layout such that it may not be easily understood by its occupants?*
- *Are there any special processes that could lead to an above-average threat of fire?*
- *Are there any circumstances where the application of management, systems, and procedures could be open to misinterpretation?*
- *Is the relationship between fire safety management and other building functions unclear?*

If the answer to any of the above is *yes*, then it is possible that the preparation of a fire strategy is warranted. Other drivers for a fire strategy may be the need to meet legislation, building control requirements, or requirements from insurers. This will be considered later in this book.

The title of this chapter points to the need for the fire strategy to be strategic, and this is the whole point of the book. The person or persons tasked with preparing a fire strategy, subsequently referred to as *fire strategists*, will need to think in a strategic way. Think more like a chess player than a technician!

It all starts with asking fundamental questions; those beginning with why, how, when, what…

Many who work in the fire industry have progressed in their career by learning *real* fire safety engineering mostly on the job and by picking up from others who are perceived as experts. Educational courses are widely available and play a big part, but what we all discover after spending many years in the industry is just how wide and encompassing the subject is. Many of us tend to specialise in certain aspects of a fire strategy, such as fire compartmentation, *or* fire risk assessments, *or* smoke extract systems, *or* fire suppression systems. So when we are presented with the task of preparing a fire strategy, surprise, surprise, passive fire protection specialists will concentrate on a strategy making extensive use of fire compartmentation; sprinkler advocates will provide strategies making use of suppression systems; fire detection experts will... You get the picture.

We all tend to focus on areas where our knowledge and experience are strongest, so the eventual fire strategy will be one that reflects our own skills. As a consequence, the strategy will be narrow and non-strategic.

Then there are the fire strategists who have successfully prepared strategies for years and take their ideas from one building to the next, even if their last strategy was for a hospital and the next is a railway station. The fire strategist has the answer; all he or she has to do is to make the building fit with the strategy. I call this approach "cut and paste" fire strategy making. Again, anything but strategic!

To be strategic, we always need to start by being objective and ask the fundamental why, what, how, and when questions.

Many years ago, a British Standard (BS 7974) was published that introduced the concept of a qualitative design review (QDR). The review process was, and is, deemed to be a fundamental requirement, especially if a fire safety engineered "performance-based" solution is used.

The QDR should incorporate the input and views of stakeholders involved or affected by the fire strategy. As well as the fire strategist(s), building representatives, fire-enforcing authorities, architects, project engineers, insurers, and building control and approval authorities may be involved.

The QDR will help identify the objectives for the fire safety engineered design (Life, Property, Business, and the Environment). It will help identify the characteristics of the building, its process and occupancy types. It will help identify the key risks and hazards and help determine likely fire scenarios and design parameters.

One of the recommended outcomes of the QDR process is to establish acceptance criteria, highly necessary if the design is not following a prescriptive set of standards. Such criteria may include evacuation times, tenability limits, structural requirements, and risk mitigation requirements.

This process forces a holistic approach and the consideration of all options, not just those that are most familiar to the fire strategist(s). The BS 7974 QDR process also prompts consideration of quantification methods, requiring the fire strategist to determine the most appropriate method to substantiate their design. The three methodologies given are:

Comparative study: Comparing the fire safety building design to equivalent or near-equivalent building designs that comply with prescriptive standards

Deterministic study: Assessing whether a predetermined condition will or will not occur based on worst-case scenario assumptions.

Probabilistic study: Assessing the probability of certain scenarios and whether they should be subject to further analysis.

Each of the above promotes thinking before doing, which, in turn, contributes to a holistic approach to the fire safety design of a building. As the QDR process is recommended at the front end of any building or infrastructure project, it should ensure that the eventual strategy covers the key issues, as well as includes the requirements and thoughts of each of the stakeholders. The QDR process promotes strategic thinking.

I started this chapter by giving the idea that fire strategy making could learn lessons from those who prepare strategies for military or business use. It is the military version that I believe is most relevant, and so the remainder of this book takes the military theme, as our armies, navies, and air forces around the world use the concept of strategic thinking day in, day out. Like the best military and business leaders, the fire strategist needs to take a step back to appreciate what *really* needs to be achieved.

CHAPTER 2.
OUR ENTENTE CORDIALE: TEAMWORK.

"Teamwork is the ability to work together toward a common vision. The ability to direct individual accomplishments toward organizational objectives. It is the fuel that allows common people to attain uncommon results."
—Andrew Carnegie, steel tycoon and philanthropist

Caroline has worked in fire safety for just over ten years. In that time, she progressed from graduate engineer to the title of principal fire engineer. She had been involved in a range of projects over the years, taking increasing levels of responsibility as she went. Then, along came an exciting project: a fantastic newly built, high-rise office complex in the Middle East.

Caroline would be the lead fire engineer, and she had a junior engineer with her. She was so excited at the prospect of leaving a freezing Chicago to spend months in the sun of the United Arab Emirates.

Forty-eight hours after she arrived, she was whisked to her first project meeting. Caroline was taken aback by the sixteen or more people around the table. "So many!" she thought to herself. She was also struck by the two-man delegation representing the *architect—but not in a positive way. During the introductions and early deliberations, she felt hostility towards her and her colleague. As the discussions went into detail, counterviews were raised to her comments at every juncture.*

By the third meeting, Caroline felt that the time had come to confront the architect's team.

Over coffee, she asked them a straight question: "Is there a problem?" Over the course of the next twenty minutes, it was revealed that there was indeed a problem, not with Caroline but with fire engineers in general. The architects had worked alongside fire engineers in the past and had formed a deep mistrust of them. Reasons given included:

- *Depending on who was involved, they seemed to get different levels of advice, sometimes even by different persons in the same organization. Sometimes the advice conflicted with what they were told by other fire consultancy practices.*
- *Most of the advice appeared to be in the form of opinion rather than cold facts. How could they be sure they what they were being told was correct?*
- *Sometimes, they found fire engineers inflexible to new ideas, and this caused much frustration.*
- *They often believed that they were getting poor value for money for the amount of input given by fire engineers.*

Caroline also learned that in some cases, they had simply stopped using fire safety engineers and used the fire safety codes themselves to ensure that compliance was covered. At this point, she had two choices. The first was to accept her lot in the project and simply undertake her role the best she could.

OUR ENTENTE CORDIALE: TEAMWORK.

The other was to change the mind-set of the architects and other stakeholders, to show the value fire safety engineering can bring to the project. She chose the latter.

Over the coming months, Caroline showed the future building owners that by considering business continuity as part of the fire strategy, they need not also develop a complex crisis management system. She proved to the construction team that they could reduce the levels of fire cladding on some of the vertical steel members that they had assumed would be required. She set up a series of fire and evacuation models to show the architects how subtle changes in the layout could bring positive results for the fire strategy. She also won the trust of the civil defence authorities by presenting the strategy in a structured and logical way.

At the end of the design phase of the project, Caroline gained a great deal of respect from the stakeholder team. She even received a good bottle of champagne from the architects!

At the same time in Amsterdam, Hugo was pleased with his latest commission to formulate a fire strategy for an airside building at a major international airport. The building was designed to be state-of-the-art and combined a number of functions all served by a novel central atrium. Hugo loved challenges such as this, and his thirty years in the industry meant that there would be nothing that would faze him.

He wasn't keen on some of the modern building designs and believed that architects never considered the needs of fire safety and protection the way they should. He found that project managers were always trying to cut costs and always saw fire safety as an easy area to achieve this. He tended to side with the enforcers, particularly because he was from an enforcement background himself.

At the first project meeting, Hugo determined that the best way to ensure that fire safety was treated properly was to stamp his authority on the team. This he did, and when some new ideas were put on the table, he reminded the team that if they really wanted the scheme to be accepted, they really should do what he said. After all, he was the fire expert.

During the meetings that followed, concerns were raised as to whether some of the fire safety provisions were strictly necessary. Hugo would have none of it. The scheme and the accompanying fire strategy were finally approved by all stakeholders, and Hugo was proud of another job well done.

It was almost a year later when the designer and installer of the smoke extract systems came to the project manager with some serious queries. "Look" they said, "we have installed smoke extract systems in all types of building, all over the world, and have never come across something so large and complex before, for a relatively small building. You could extract smoke generated by a fire five times the size of the maximum one you will ever get here, and you would still have capacity for more."

The project manager looked again at the plans for the smoke extract arrangement and questioned why the systems were a sizeable part of the overall budget. He called in another fire consultant he had used in the past and, sure enough, an appropriate redesign was submitted, just in time for the construction phase! There were other queries raised, such as the necessity for two-hour fire doors for the external walls of the building."

The above two fictional examples could be regarded as extreme cases of two approaches when working with a project team.

OUR ENTENTE CORDIALE: TEAMWORK.

The message here is simple. It is not just about education, knowledge, and experience. Attitude plays an important part in the eventual success of the fire strategy.

Ideally, the fire strategist is involved right from the early feasibility stages of a project. It is the fire strategist who can add value by ensuring that building design development progresses in the correct manner and that those with an interest that the building is fire compliant are consulted regularly, and that comments are properly taken on board. In this way, there will be no surprises and no hasty redesigns just before the construction stages. If only every project went this way!

In chapter 6, I describe fire as the enemy. Even though we understand the fundamentals of fire science, we also recognise that a real fire in a real building does not always conform to the understood science. This can sometimes lead to alternative views as to the most appropriate fire strategy. It can involve a large element of "opinion engineering." Although experts in their field are entitled to their viewpoint, the danger is that where agreements are required, the solution tends to be based on those with the loudest voice rather than those with the best solution. This is a trait that goes against the concept of thinking strategically.

> **Fire Strategists: How do you manage your role?**
>
> **Others: How do you view the role of fire strategist?**

Even the best military leaders would have surrounded themselves with advisors, and, in the same way, the fire strategist should be open-minded to the ideas of others.

I hinted at one of my key concerns in the Preface—that fire strategists are not as valued as they should be—and this is borne out by the experience of many fire strategists around the world. Why is it that, in a number of projects, fire specialists are brought into a project as a last resort or when another stakeholder raises an issue that they cannot address? Why isn't fire expertise always deemed fundamental from the beginning of a project?

Could the stories behind Caroline and Hugo give some clue to this?

Who is a fire strategist?

The profession of fire engineering or as it is also known, fire safety engineering, has greatly changed over the last couple of decades, particularly due to the fact that performance-based solutions have become more mainstream. This requires more in the way of training and knowledge than would be necessary for adopting a purely prescriptive approach. Being able to recite key paragraphs and clauses from a range of codes may have been the prerequisite for the fire strategist at one time, but today, a scientific and objective mind is necessary. Put it another way, the barriers to entry into professional fire engineering have been raised by the requirement for additional knowledge, experience, and, of course, attitude.

OUR ENTENTE CORDIALE: TEAMWORK.

The determination of the right level of competency to prepare, or even be involved in, a fire strategy, is still a long way off from resolution. Those who are probably not as competent as they should be are those who may impact the way all of us are seen to the outside world. This issue is not just found where fire engineering consultancies offer their services to an external market, but even when the fire engineering business is part of a multidisciplinary organization incorporating architects, civil engineers, and so on.

Let us look at the profession from the eyes of others. We all know deep down how the "fire engineering" experience varies greatly across the board. Even within single organizations, depending on who is assigned to the project, a totally different outcome can result. Some consultants may come across as helpful and others quite belligerent, that is, "If you do not do as I say, I will not guarantee that this will get past building control." This variable experience can set alarm bells ringing to those who want to, or are forced to, use professional fire engineers.

Another issue is how we deliver the fire strategy. The final fire strategy document can vary from what would appear to be a generic specification, with or without drawings, through to complex reports encompassing detailed analysis. What a fire strategy should and should not include is often in the mind of the person preparing it. This is one of the reasons why I thought it would be a good idea to standardize the process with a Standard: BS PAS 911.

So how can we assure ourselves that persons preparing the fire strategy are competent and capable of undertaking the key tasks? Perhaps some measure can be taken from recent British legislation when it concerns the undertaking of fire risk assessments.

The UK Regulatory Reform (Fire Safety) Order requires that a competent person or persons is/are employed to assist a person with designated responsibilities for meeting the requirements of the law. It defines a competent person as "a person with enough training, experience or knowledge and other qualities to enable them to properly assist in undertaking the preventive and protective measures." Although this definition is for a fire risk assessor and not a fire strategist, it clearly does not take away ambiguity.

Defining a competent person is far more complex than simply measuring relevant experience or qualifications. In order to understand *real* competency, perhaps we should look for clues by understanding how persons enter the profession and become fire strategists.

There are a number of routes. Here are some of the most prominent:

Firefighters: Persons who join the fire brigade or civil defence authorities and then move into the capacity of professional fire engineer. They may have had a career in mainstream firefighting or in fire safety and decided to be more actively involved in fire engineering. The benefit of having ex-firefighters is that they understand the impact of fire from direct experience.

> **How risk-averse are you?**

They may know what works and what does not. They may have seen a variety of fire strategies, whether implicit or explicit. They may have moved into enforcement, where they have been given the authority to say "yes" or "no" to proposed fire strategies. Their background may, in some cases, also lead them to be more risk-averse and to be cynical about new ideas or techniques, based on their direct experience of many facets of the fire industry. The ability to question new ideas can often be a good thing, as it can sometimes balance the views of other professional types, particularly academics.

Academics: Persons who study the subject or are involved in research and development. In this case, I would not define academics as those with a first degree in fire engineering but those who have at least studied to masters level and/or have been involved in research. Academics are a fundamental part of the fire industry when it comes to developing and adopting a performance-based approach. In many ways, academics make the most natural fire strategists in that they are taught to be open-minded to alternative ideas. They are objective and holistic thinkers in general, and they can adopt a controlled methodology in approach that could capture all the key issues succinctly. What they may lack, however, is the knowledge of the practical application of their ideas and an understanding of the consequences of their solutions. Furthermore, their goal may be to push back the boundaries of current knowledge to come up with new and radical ideas; they are likely to be more risk-taking than risk-averse.

Postgraduates: Those who have studied a first degree in fire engineering or a related subject and have entered the profession at a relatively early age. Their views, risk profile, and experience will largely be moulded by the persons around them, particularly in the first five years or so.

Engineers: Persons who enter the subject from a core engineering subject such as construction and electrical or mechanical systems, or directly from fire protection engineering companies. They may be involved in designing systems and/or components of systems or may be involved in specialist engineering applications. Such persons will, in general, be adept in finding practical solutions and will introduce a common-sense approach. They will probably be best utilised to develop tactics for the delivery of the fire strategy than to develop the strategy itself.

Risk assessors: Persons who enter the industry via a background in assessing risk. These may come via the insurance route or via related sectors such as health and safety. They may also have been involved as enforcers and have moved into risk assessment. As with engineers, risk assessors will have a practical and "hands on" approach to the key issues. They may be less likely to be comfortable with the more ethereal aspects of performance-based fire safety engineered solutions but will bring another practical dimension to determine appropriate conditions leading to a fire scenario. Bearing in mind that they will be practitioners, they may also tend to be risk-averse and cautious about non-standard ideas and applications.

I have been very general in providing "typical" profiles. Let us not forget that, even within each of the sectors, similarly experienced persons may see things quite differently. But adding the skill sets and risk profiles together could prove to be very powerful in providing effective and realistic fire strategies.

Opinion engineering, black art, beautiful subject—it is not surprising that fire engineering can be viewed in so many different ways given that each of us in the industry is likely to view the subject in a slightly different way. Given this, is it possible to say whether one person is competent and another is not?

Who is more competent to prepare a fire strategy: a fireman who has spent twenty years fighting fires and knows, first hand, the real issues, or someone who has designed fire sprinkler systems for fifteen years and has helped write sprinkler standards?

Who is more competent: a scientist with a PhD in a branch of fire safety who has published several research papers over twelve years, or someone who has spent his or her career assessing the fire safety risk in a range of sites from power stations to hospitals?

The subject of competency cannot easily be resolved, but there is common agreement that a combination of relevant experience and professional qualification or affiliation is a good basis.

The UK-based Institution of Fire Engineers has, in recent years, introduced an Engineering Council, which has paved the way for Chartered Engineer Status not just in the UK, but around the world. This is equivalent to the US Professional Engineer (PE) status. Persons qualifying as a chartered engineer will need to have gone through a rigorous assessment by the Institution, which considers both qualifications and experience (and, to some degree, attitude). Engineers chartered by the Institution could therefore be regarded as competent persons. This being said, there are probably many worthy, competent persons without this professional qualification. Life is never straightforward!

Many professional fire advisory and consultancy companies have persons who derive from different backgrounds; firefighters, engineers, academics, risk assessors and so on. It is this mix that may be most appropriate for strategic thinking, as it encourages a cross-fertilization of ideas and a questioning of initial conclusions. Only in this way do I believe that a robust and long-lasting fire strategy can be produced.

A strategy prepared by a single person, however competent, will be subjective. A strategy prepared by a team is more likely to be objective, and any holes in any one person's knowledge and experience are likely to be filled by others. There may be a case for bringing in persons from other "competitor" organisations to assist with more complex scenarios. Perhaps this is a habit that fire strategists should consider. Peer review can be very effective. Serious thought should be given to fire strategy preparation by a team rather than an individual.

Bearing all this in mind, perhaps we should look at a competent person in a completely different way. Possibly, a competent person is someone who knows when they are *not* competent to undertake a task, someone who says "No, I don't have the level of experience or knowledge to properly advise on that."

Competent persons should not prepare a fire strategy without consultation. At least, they shouldn't prepare the strategy without consulting relevant stakeholders or other interested parties. Consultation of relevant bodies is highlighted in any number of standards and guidance yet is not necessarily followed in the spirit of the recommendation. There may be very good reasons that restrict participation, but it is the responsibility of these competent persons to ensure that all relevant persons have been invited. This is not just for the inclusion of alternative ideas or to ensure that all bases have been covered. It is to ensure that once the strategy is complete, all relevant parties do take ownership of it, as they have played a key part in its preparation.

The Fire Strategist as Part of the Team

When we talk about team involvement in the preparation of a fire strategy, we do not just mean the fire strategists working on the project, but everyone who plays a part in its development. This could include the building owner/operator, project management, architects, building control, enforcers, insurers, and possibly other interest groups. Getting the key people on board at the early stages of fire strategy preparation is key to its success—an idea that sometimes falls on the wayside as the project gets going and we all get caught up in the project momentum. Typically, a number of bodies could be part of the team:

- *The person(s) preparing the strategy, i.e. the fire strategist(s).*
- *The end client, or those representing the fire safety interests of the building or organization. Note that this could include appointed project managers.*
- *Architects and those involved with the design of the building.*
- *Enforcing authorities, including fire and civil defence authorities, building control, etc.*
- *Insurers or those required to ensure asset protection, business continuity, etc.*
- *Sector-based authorities: Persons representing other building or organisation interests affected by the fire strategy (e.g. heritage organisations, government inspectorates, facilities management, etc.)*
- *Those involved with, or who will be involved with, the day-to-day fire safety aspects of the building. This could include fire systems servicing engineers.*

When we all work together to ensure all viewpoints have been properly taken into account, the project can be a pleasure to be a part of. However, as budgets change and milestones get missed, there may be a time when "entente cordiale" becomes anything but.

We should never forget that every stakeholder in the team will have a view and will want his or her ideas to be taken on board. We will start with the end client—the building owner or agent or the company whose infrastructure is the subject of the fire strategy. In their case, they will want the fire strategy to be as effortless as possible and would like to see one that ensures that they fully comply with minimal outlay or disruption. They are unlikely to appreciate the full impact or benefit of fire safety and protection systems other than that they will ensure that they comply with the law of the land.

The fire strategist can play a vital role in educating the client to the issues other than the provision of fire compliance. In chapter 3, we consider the greater objectives that could be incorporated into a business strategy, including property protection, business continuity, and protection of the environment. A fire strategy can unify a number of objectives and provide a solution that covers them all rather than treating them as separate issues.

We must not forget that the client is probably not meant to be clued up in the subject of fire safety, and when the fire strategist makes proposals, they need to be put across in a way that is clear and unambiguous. Another faux pas is when a fire strategist makes a proposal and then asks the end client for his or her opinion or approval.

The equivalent situation would be a plumber coming to your house and asking your opinion on how he should fix the boiler. You would be bemused and question his ability. In the same way, seeking approval for an element of a fire strategy from a non-fire professional is not appropriate. Note that the client may be represented by other appointed bodies, such as a project management team, who should be treated with the same amount of understanding and courtesy as if they were the client themselves. Never forget that the client (or the client's representatives) utilises fire strategists for their expertise and should not be brought into the technical consideration, unless they choose to be.

The next group who may be part of the strategy team is the architects. Their objective may be to push back the boundaries of building design, which can be at odds with the aims of the fire strategist. For the more innovative projects, following prescriptive fire safety guidance, largely based on "typical" building profiles, could be a problem. Performance-based fire safety engineering has evolved to keep pace with modern building design. Going back to first principles and determining what the performance objectives are should be seen as a great boost by architects. But this is not always the case.

There are architect practices that believe they can do without the service of fire engineers. They may choose to apply fire safety guidance as given in building regulations, themselves, possibly unaware of the benefits of the performance-based approach. Fire strategists often have the opportunity to sell how their involvement will cut through so many of the issues many architects are just not aware of. Fire strategists can, and should, ensure that they work alongside the architect and are seen as a team colleague rather than an adversary.

The enforcer may or may not be a key part of the team but can, and often will, be involved in the final decision-making. I would describe enforcers as any party that can use legal or regulatory powers to influence the outcome. This will include fire and civil defence authorities, building control authorities, and, in some cases, the insurer. It is vital that the fire strategist keep them alongside from the start of the strategy process, especially as they have the power to overturn or at least question any and every aspect of the strategy, sometimes at the least opportune moments.

If the strategy leans towards a prescriptive approach, then the task is likely to be substantially easier. If a performance-based approach is used, then the criteria and scenarios proposed at the early stages must be also put to the enforcers for their agreement. Even if proper reasoning has been used by the fire strategist and it has been documented accordingly, there is still the potential that the trail of decision-making is not robust enough.

It should also be remembered that enforcers have key responsibilities to the community to uphold minimum standards of fire safety. They are unlikely to agree to arbitrary decision-making. Consequently, in order to keep enforcers happy, they should be consulted every step of the way

I mentioned the insurer involvement in the process. My background and my introduction into the fire industry were in representing the fire insurance industry. In many ways, fire insurers have been the instigators of modern international fire safety and protection practice, through their need to protect their investment. The idea of professional fire brigades was largely influenced by fire insurers, with the primary aim to put fires out in buildings insured by them. Some of the original fire codes were developed by them, and they played a key role in focusing the development of passive and active fire protection. Today, their influence is widespread, although their involvement has somewhat diminished. The fire strategist should, as part of the strategy process, check whether there are any special requirements dictated by insurers even if the insurers themselves are not obvious in their attendance.

Other parties involved may be sector-based authorities. I have worked with a few of these during my career, and they can offer valuable insight into the practical issues affected by the strategy and can also provide solutions to seemingly problematic issues. Once again, involving them at the very early stages of the project can reap rewards further on. Examples of sector-based authorities will include heritage and conservation groups, sector representative groups, and government-appointed inspection bodies.

Once the team works together, and their distinct roles and responsibilities are understood by all, the development of the strategy can be made easier.

One way to formalize the strategy preparation process is by forming a panel that should meet at key stages of the strategy development. These stages may include:

- *Planning meeting(s)*
- *Meeting(s) to consider specific issues, such as objectives setting, risk assessments, and building issues.*
- *Technical meetings to consider aspects such as system technologies and construction methods.*
- *Review meetings at key milestones to review work stages.*

Although it may be desirable that all members of the team attend every meeting, it is practically unlikely. Consequently, I would recommend that at least three milestone/review meetings be held. The first should occur at the early stages of preparation and after the fire strategist has undertaken an initial review of all the key issues that may have an impact on the strategy. Earlier in this book, I identified a type of review called a "qualitative design review" or QDR. This is a structured approach to an early meeting with all members of the team and will help identify all the key issues.

The second review meeting should be held after a draft fire strategy has been prepared. When members of the team start to see the strategy as it is presented, the opportunity to amend will be there. If the original meeting was thorough enough, the second review should ideally be confirmation that the strategy has been developed in the right way. A third review meeting should be held after the draft strategy has been completed to ensure that it takes into account every aspect that has been raised. This is also an opportunity to test the robustness of the strategy.

This chapter has been more about the human dynamics of strategy preparation. I am not saying that the fire strategist should be a psychologist, but it is the interaction between team members that will play a large and crucial part in the success of the fire strategy. It is not just a case of strategic thinking; it should incorporate strategic working.

There is a lot more to getting a fire strategy right than simply possessing qualifications and experience. As with everything in life, attitude plays a key role, and, in order to end up with an effective and robust strategy, we must be inclusive of all stakeholders and work as a team. In particular:

- *The fire strategist(s) must ensure that ignorance plays no part in fire strategy preparation. As the saying goes, "If in doubt, find it out!" Arrogance and single-mindedness are also traits that will prevent strategic thinking by the team.*
- *It is up to the fire strategist's team to make sure that the experience as part of the team is a positive one; we are there to help rather than to hinder.*
- *The end client (or the client's representatives) utilises fire strategists for their expertise and should not, implicitly or explicitly, be given responsibility for making technical decisions, unless they choose to be.*
- *Fire strategists can, and should, ensure that they work alongside the architect and are seen as a team colleague rather than an adversary.*
- *The fire strategy process should be robust enough to retrospectively identify the background of decisions made.*

CHAPTER 3.
THE MISSION

"If you find yourself in a fair fight, you didn't plan your mission properly."
—David Hackworth (Colonel, US Army)

Every form of strategy, whether it is for military purposes, business development, or fire safety, needs to have a clear idea of what it is trying to achieve. This may seem an obvious idea; nevertheless, thinking about the purpose of the fire strategy, as with any strategy, is a key driver in deriving the right answer.

Determination of one or more key objectives will be the focal point for all that is to be subsequently undertaken. The term "mission," deriving from its military roots, was a buzzword in international business from the 1980s, when every company had to have a "mission statement"—a sentence or paragraph that succinctly states why they do what they do.

It is not uncommon in the world of fire safety and protection for assumptions to be made as to why a set of fire precautions are deemed necessary. It will be assumed that fire safety is a fundamental requirement for every building and that the prime purpose is to meet with national legislation. It will be assumed that everybody, including fire strategists and other stakeholders, understands the need for good fire safety management and for proper levels of fire protection. It may be assumed that, by following a process without question, we will arrive at the right answer. Wrong.

Imagine the setting when the fire strategists, architects, enforcers, project engineers, and so on are sitting around the meeting table. Imagine that the first question coming from the fire strategist is "Why are we doing this?" The imaginary response may be "Well, if you don't know, how the hell are we supposed to know?" But the question is valid.

A fire strategy may be required to ensure that the building or infrastructure meets the relevant requirements of national rules or regulations. But there is so much more that a fire strategy can help with, potentially providing so much more value than a strategy simply designed to ensure legal compliance. The remainder of this chapter provides some thoughts on this.

As part of the fire strategy process, fire strategists will have the opportunity to delve into the thinking of all the parties sitting around the table. They have the potential to turn a group of "adversaries" into a team of "allies," all working towards a common purpose. After all, the other stakeholders around the meeting table are not the enemy; fire is!

This can all start by asking a simple and apparently obvious question. We really do need to ask the question "why," even if we think we know the answer. As with many things in life, the answer is never so straightforward. If the "why" question is asked at the front end of any strategy preparation, then the amount of resources wasted by revisiting issues as they come along will be reduced. In fact, a good bit of advice once given to me is to ask the "why" question three times, to get to the crux of any issue.

The first why: Asks a simple question and in the majority of cases a simple and seemingly obvious response will be given.

The second why: By this time, you are in fact questioning the supposition that may have simply been a stock response. At this stage, some thought will need to be given to the response.

The third why: By this time, serious thought does need to be given by the responder. This is where the retort of "I don't know" or "I'll have to get back to you" may come in. If there is a valid and lucid reply by this time, then the responder must know what he/she is talking about.

The standard or stock response for the objective behind the fire strategy is to achieve "compliance." This normally implies compliance with national legislation. I refer to legislation as part of the mandatory framework, requirements that *must* be incorporated into any fire strategy. Or must they? Let us first examine how legislation in the field of fire safety has typically developed.

At some point in history, there was deemed a need to legislate minimum requirements for fire safety, on a regional or national level. This was often a result of a major fire or acknowledgement that without such legislation, the trend in, say, fire deaths will be

FIRE STRATEGIES - STRATEGIC THINKING

unacceptable. Sometimes legislation was rushed in to fill a void in regulation. One such example is that of a piece of British legislation titled the "Fire Precautions (Sub-surface Railway) Regulations 1989."

Following the London Underground Kings' Cross fire in 1987, when thirty-one people died, there was a big outcry by the general public and by government. Something had to be done, and done quickly (no doubt to safeguard the concerned passengers). A research study was undertaken, and the possible causes of the fire were determined. It was these causes that led to the drafting of specific requirements that formed the basis of the Regulations, which became law within two years of the fire itself. Within the Regulations was a list of prescriptive measures, including:

- *The use and arrangement of electromagnetic or electromechanical devices to hold open fire doors.*
- *The type of suppression system required for protection of escalators and travolators.*
- *The requirement for linear heat detection systems to monitor escalators and travolators.*

The question of how far national legislation should go in prescribing fire protection is a valid one. Does not the inclusion of statements, such as the examples given above, stifle technical innovation if they are moulded into policy?

> **The fire strategist should question everything!**

The lesson here for the fire strategist is one made throughout this chapter: it is right to question everything, even if it appears to be set in stone!

Normally, legislation is supported by codes and regulations. This is obviously preferable to having them embedded within the words of legislation itself. The words behind any piece of legislation are written by well-intentioned people, either singly or collectively, and sometimes written during times of emotion. They will then be subject to scrutiny and will be modified to cater for the range of views. It is quite possible that somewhere down the line, changes in the meaning are made to provide compromise. The end result may not provide the intended solution. Sometimes the written law could well be fallible.

Consequently, if a fire strategist is faced with a regulation that is impractical or just not possible for that set of circumstances, he or she should have every right to question it. The building, its processes, and its people, or as I later describe them the "terrain," may not conveniently fit into the box marked "legislation." It is far better to have an effective fire strategy than one that has its corners cut to fit. As far as a fire strategy is concerned, *everything* can and should be questioned.

The mandatory framework may also include the requirements of insurers. At one time, insurers were heavily involved in the risk nature of a building. It was the insurance industry that introduced professional fire brigades. It was the insurers who developed many of the early codes around the world and who played a part in moulding the way fire protection has been developed to date. In some parts of the world, the fire insurance industry was extremely powerful. In recent times, the insurer has, in general, become less proactive, largely due to the forces of competition and the desire to cut costs. Nevertheless, insurers may have a part to play in the strategy, particularly where there are property protection or business continuity issues involved. Whereas legislation is typically black-and-white, the insurer may be more pragmatic and may even offer solutions that can be directly incorporated into the strategy.

The mandatory framework may also need to include the requirements of specialist groups. Examples of such groups include heritage bodies, whose role is to ensure that whatever the fire strategy dictates, it will not adversely impact the fabric or aesthetics of the building. Transportation inspectorates will want to ensure that the fire strategy caters for their requirements for both passenger safety and for the running of the networks. Most sectors will include specialist groups who may all have an impact on the objectives of the fire strategy. It is up to the fire strategist to ensure that all relevant organisations are involved in the objectives of the fire strategy. This is why getting the right team together and working together early on in the process is the key to success.

What If?

Putting the mandatory framework to one side for a moment, let us imagine a world of no fire safety legislation, no building control, no insurers, and no specialist groups. In effect, no actual requirements to do anything at all with respect to fire safety and protection. The big question is, would there be a fire industry?

Would building owners choose to install fire detection and alarm systems for their own ethical reasons? Would they protect an escape staircase with costly fire doors and pressurization systems if they didn't have to? Would the architect feel duty-bound to consider the dynamics of fire movement when they are designing the next most elaborate construction to outdo their competitor?

> *Imagine a world without fire safety rules...what would be our role?*

Fire strategists must keep this question in the back of their mind; otherwise, they are simply servants of the mandatory framework. What we need to do is ask the question, why (times three) are we doing this? Let us go back to basics.

None of us would really enter a building if we knew it was a fire trap. We would not stay in a hotel if we were aware that there were no safety systems in place to protect us should a fire start in one of the rooms.

Would any of us travel up the escalator to the fifteenth floor of a high-rise office block if we thought that we would not be able to get out again if a fire burned on a floor below?

This is where our internal fire strategy kicks in. Even in a world without mandatory fire safety, our need for life preservation would hope that the building owner

FIRE STRATEGIES - STRATEGIC THINKING

or operator has taken the ethical decision to protect us. The same ethical issue applies to property protection and business continuity—we would like to think that the factory in which we work would not completely burn down if a fire started, and that we would still be able to work and collect our salary. Then there is the question of protecting our environment for now and in the future. Sometimes the chain of events can lead to unrealised results, pushing the decision to incorporate fire precautions towards ethical and commercial considerations rather than purely legislative ones. Let me give you two fictional scenarios based, somewhat, on actual events.

Scenario One concerns a newly opened airport terminal on the beautiful small island of Shonata.

Shonata is a picture-perfect retreat with white sands, aquamarine seas, and a lovely all-year-round climate. It has become a tourist hotspot, creating the need for a brand new Terminal 2. On the day of opening, there was still a rush to complete the various installation work, but the opening day could not be delayed due to the visiting dignitaries. The contractors were given the day off. They could complete the installation after the opening ceremony.

As well as the gorgeous weather and beaches, Shonata was famous for its food, especially its marinated and deep-fried fish and meat. A brand new "Shonata Cuisine" restaurant was incorporated into the terminal, and the dignitaries would be the first to try the restaurant cuisine. So, prior to the visit, the head chef lit the burners to heat the oil. He was unaware that the cooking equipment had not been properly tested. The oil caught light and the flames burst forth. Unfortunately, the cooker hood fire suppression system had not been commis-

sioned, so within seconds the fire took hold. None of the restaurant staff had been fire safety trained. That was planned for the following week.

The chef instinctively took a pan full of water and threw it over the burning oil. He was engulfed by flames and died painfully. Unfortunately, the fire doors and separations designed to contain the fire (although these were subsequently shown to be totally inadequate) had not been properly snagged. Within two minutes, the fire had taken hold of one whole section of the new terminal and claimed another two lives, as well as putting thirty-five into the hospital. The aftermath was measured as follows:

- Fines and costs relating to the dead and injured: $12 million US.
- The cost of rebuilding the Terminal: $35 million US.
- The costs estimated in loss of tourism and bad publicity associated with the fire: $250 million US.

Without even considering legislation and building control, the cost associated with the loss of life would be regrettable, but it is the property costs and most significantly, the business costs, that would cripple Shonata's economy for years to come.

The next scenario takes place in rural England.

John was a farmer of arable land for many years before retiring. In the last few years, due to bad health, he let his land and outbuildings go to ruin. He was making some money letting out his barns for storage and other uses to local businesses. John's farm was just outside a pretty village called Bottom's Edge, loved by professionals due to its relative proximity to London.

Without John's knowledge, some local youths regularly used one of the barns as a den. One dark night, as they sat *around smoking weed and drinking cheap vodka, one of them thought he heard a police car. They panicked, spilt the vodka, threw down their "cigarettes," and ran. The combination of the discarded materials ignited a fire that quickly grew. There was no fire detection or protection of any kind, so the fire had time to grow until the whole barn was engulfed. The flames and the huge smoke plume in the night sky were first noticed by the villagers, especially when the cooled plume started to settle in the village. The fire brigade were called, but by the time they arrived, there was very little to salvage.*

Fortunately, nobody was hurt, and the redundant barn had very little property value and was simply used to store waste. It was subsequently found that one of the businesses using the barn was an asbestos contractor and had used the barn to store waste material rather than to dispose of the material properly. This information got out, and one of the villagers, a tabloid newspaper reporter, realised that the dust and film settling onto the village, and especially the school playground, was not as harmless as they had been led to believe.

John and his business friend are now fighting litigation on the grounds of exposing the local environment, and especially the local children, to a health risk. The potential for compensation could be enormous! No life safety issue, no property protection issue, no business continuity issue, but one massive environmental issue.

This is the point of this chapter. The fire strategist has an opportunity to ensure that the strategy takes into account all relevant issues, many of which the client may not be aware of, but may be thankful of if the extent of their risk is really known. By examining the needs of the building, its owners, its occupants, and its processes, there may be a number of objectives relevant to the fire strategy over and above the requirements of the mandatory framework.

Figure 1 shows four key objectives: life safety, property protection, business protection and continuity, and environmental protection. Each of these objectives is broken down into four sub-objectives, as described below.

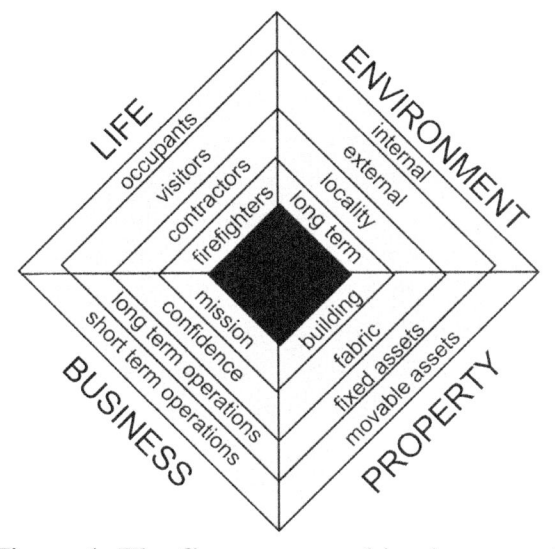

Figure 1: The fire strategy objectives matrix

Life Safety

Let us first consider the key objectives of life safety and in particular the building occupants. Fire safety legislation is designed to ensure that the normal occupants of a building are, as far as possible, kept safe from a fire, and when required, can evacuate safely. Typically, legislation focuses on the life safety of occupants of the building. Consequently, by meeting legislation, you could assume that the life safety of the occupants of the building has been covered, which in many cases may be correct. Even where prescription is used, the fire strategist should possibly look deeper and consider the profile of occupants, their numbers, the level of disabilities, cultural or language issues, etc. Occupant profiles are covered in more detail as part of our assessment of the "terrain," as described later in this book.

The second sub-objective is visitors. Visitors may have a different profile from occupants. In this case, the fire strategist should assume that they have little knowledge of the building, and their requirements for being in the building may be varied. In many cases, visitors to a building may be a small proportion of the overall occupancy, such as in an office block. The general public will also count as visitors, so a railway station, airport, museum, or sports stadium will have the majority population as visitors, with all their various profiles. By considering visitors separately from occupants, the fire strategist should utilise different assessment criteria.

Whereas standard occupancy numbers are normally known with some accuracy, the numbers of visitors may change at any time of the day. The fire strategist may choose the maximum possible numbers to work with, or may use statistical data. Visitors may also be unevenly spread out across the building; they may exist in high concentrations in specific parts of the building only.

The third sub-objective is contractors. In this case, contractors are those persons who work on or in the building, whether for maintenance purposes or in construction projects. But why treat these persons as a separate category? How many fire strategies consider contractors differently from other building occupants?

The main reason for inclusion as a sub-objective is that they will normally have different characteristics from occupants. They may or may not know the layout of the buildings. Their numbers may vary over the course of a day or a week. However, the main reason for their inclusion has to do with evacuation times. When we normally consider evacuation, we calculate the travel time as the time from the seemingly worst case condition to a place of safety. This could be from the far end of a single means of escape, or from an inner room, via another room, to an escape corridor. What is not usually considered is where the contractor may be when an alarm is raised. He or she may be in an attic area up a ladder, or in a confined space where movement is restricted, or even undertaking repairs to the roof.

Two things to consider here are (i) how the alarm message is conveyed in these areas and (ii) how many more seconds or minutes are necessary before the person(s) can get to a position where they, too, can

> **How long will it take to escape from a confined space in the event of fire?**

commence their evacuation. Obviously some common sense should be applied here. Taking every conceivable place where someone may be may make the strategy overly complicated. Instead, places where people could be on a fairly regular basis should be identified.

The fourth life safety sub-objective is the firefighter. Very few countries now just accept that firefighters know they are taking a risk and should be left to their own devices. If the fire strategy is purely life safety of the occupants, and there is no requirement for firefighters to assist in any way, then the safety of the firefighter can be discounted. For all other cases, firefighter safety should be considered. Anything from specific provisions for accessing every level of the building, to enhanced fire compartmentation over and above that necessary for evacuation, will need to be considered.

Property Protection

The importance placed on the protection of property from a fire will vary greatly. Whereas the main objective of a life safety fire strategy is to ensure that all persons can be evacuated safely, the main objective of a property protection fire strategy is to limit the damage caused by a fire.

This may require the summoning of professional firefighters to the scene as quickly and reliably as possible, or it may require that automatic fire suppression systems are initiated, or simply that the fire is contained by passive fire protection, or it may be a combination of tactics. Four sub-objectives are given. The first is the building itself. This requirement probably does not need explanation, and it is the building we normally think about when we discuss property protection.

The second sub-objective is the linings of the building. The most appropriate example here is that of heritage buildings. Sometimes the internal linings of the building are more valuable than the building itself. Wall and ceiling frescos are a good example of this. In such cases it is not only fire itself that the fire strategist needs to be aware of. Some frescos may be permanently damaged just from small quantities of smoke.

The third sub-objective is fixed assets—those assets that may have intrinsic value but cannot be easily moved. By treating these assets differently from other property classes, a focused approach be taken. Examples here would be computer servers, manufacturing equipment, and test equipment.

The fourth sub-objective is movable assets and can include anything from computers to works of art. In such cases, the most valuable items may justify special consideration or dedicated arrangements. What comes to mind for me is watching the news on TV in 1992 as Prince Andrew, a prominent member of our royal family, was seen carrying out valuable pictures from the burning Windsor Castle. I doubt that this arrangement was incorporated into the fire strategy.

Business Protection/Interruption

Despite worrying statistics about businesses going under after having a fire, many fire strategies just do not consider business protection or continuity. The way a business goes kaput is sometimes a gradual process that starts with the fire. Even if the employees are safe and the building is not destroyed, a business can still suffer.

In the direct aftermath of the fire, the primary focus will be on the fire and the issues surrounding the fire. The focus may be taken off the business itself. Gradually, clients may start to look elsewhere, as the level of service they enjoyed is no longer apparent. Suppliers may find that dealing with the business has become more difficult. The fickle nature of business can often lead to a slow and inevitable decline.

The initial sub-objective is to consider how a fire will affect short-term operations—today, tomorrow, and next week. Businesses may have established an effective business continuity plan that does not involve or require fire protection or fire safety management. Even so, it is worth asking the question, as some elements may not be instantly transferable and may warrant protection from a fire and from the effects of a fire.

The next sub-objective is long-term operations. Even if there is a quick fix to allow a business to survive in the short term, there may be an impact in the longer term as to how things are done. It could be easy to switch manufacturing to another plant for a week, two weeks, or a month. What about next year? Will a single fire start to affect how the business operates?

Confidence in the business as a result of a fire, or should we say, loss of confidence, can have a major impact. It is this sub-objective that can often lead to the demise of the business. Those who operate daily in the public domain, such as a mass transit operator or the owners of a football stadium, will need to ensure confidence remains high in their abilities to safeguard the public; that is, to be able to control fires with minimal loss and disruption. Asking thousands, or even millions, to use a mass transit system that appears not to be fire safe will lead to a decline in its use. However loyal a football supporter will be to their club, they still may decline from bringing their family to a place where there is even the slightest possibility of the risk of fire.

Confidence in business can be a shallow issue, but it may need to be taken seriously. In such cases a fire strategy needs to lean towards high levels of fire prevention and fire safety management instead of relying purely on fire protection.

The final sub-objective is mission, the same title as this chapter and just as relevant. If nothing else, consider how a fire can be so fundamentally damaging in so many ways that it may raise questions about the business itself. Imagine a fire safety teaching establishment suffering a major fire that destroys assets and highlights major flaws in their fire strategy. You can imagine the questions: "If these guys can't even get it right and they are training the rest of us, what hope is there for the fire industry?" Similarly, those industries that deal with explosive atmospheres or highly flammable products must take fire safety seriously. It is fundamental to their business.

The Environment

Environmental issues rarely are considered as part of the process in preparing a fire strategy. In many cases the environmental impact of a fire is not a key issue. I believe that protection of the environment will increasingly move towards part of the mandatory framework, so there is no time like now to assess this objective.

Let us start with the issue of the internal impact of a fire, that is, the impact of a fire within the building Whether the risk is from manufacturing processes or the storage of solvents or the chemical make-up of fixtures, a fire may release products that could prove a problem for the local environment. Where fire strategists see that this could be an issue, they should help clients identify how this could impact them, and they should identify prevention or limitation techniques. They also should evaluate how these products could circulate around the building environment. Note also that the products of combustion could contaminate anything within the building, such as manufacturing processes, stock, and consumables, making them unfit for use.

Then there could be issues with regard to secondary contamination, such as the pollution of water supplies. What are the immediate and longer-term health and safety aspects relevant to using the building following a fire? What costs could be associated with any clean-up operations within the building?

Similar considerations should be given to the area around the building. The external impact of a fire may well affect all neighbouring buildings. It could be smoke damage or the impact of radiative heat to buildings immediately within the vicinity. It could be damage to cars in the car parks or to neighbouring processes that may not be sealed from the impact of a fire. With the increasing appetite for litigation in an increasingly competitive and unforgiving world, the potential is there for unforeseen fines from every direction. Let us also not forget possible health and safety implications to both persons and animals directly affected by the fire.

The next sub-objective moves from the impact on the neighbours towards the locality. This is the assessment of how the region in which the building is located may be affected by the fire. Considerations may need to look at how the fire plume, if not controlled, will affect the local community. Fires that reach tens of megawatts in size and lead to the release of airborne contaminants will be a problem. This could be compounded by the impact of various weather conditions and could lead to widespread issues. It may not even be the fire itself causing the problems but the impact of fighting a fire.

> **How many fire strategies consider the environment?**

As an example of this, take the actual example of a major recycling plant in the United Kingdom. When this plant caught fire, it required huge quantities of water to control the thousands of tonnes of burning recycled wood and sawdust. The subsequent water run-off polluted a nearby canal, causing reduced oxygen levels and thus endangering the itinerant marine life. The UK's Environment Agency had to move the affected fish and other creatures to another location, a costly and time-consuming exercise borne largely by the taxpayer. How many fire strategists would, or could, have seen this cost as a result of a fire event? If they did, then they would have needed to consider the efficacy of "water run-off" containment systems. Depending on the potential quantity of water usage, a variety of means could be employed, such as bunding, creating artificial lakes or, channelling to areas that can be waterlogged without any risks to people, property, or the environment.

The final sub-objective is the longer-term impact of a fire. This may be hard to quantify and may be more subjective than the previous three sub-objectives. Nevertheless, as a community, there will be a need to at least consider certain scenarios and how they could pan out, not just in months but in years. There is little doubt that a number of major fires around the world have affected the local, regional, and national ecology. Examples that come to mind include the impact of a single fire incident on an off-shore oil platform. The release of oil as a consequence of a fire can change the local ecology for decades.

Conceivably, a fire can cause damage in so many ways. Some or most may not be relevant to the fire strategy for the building or site in question. Knowing what is relevant and what is not can turn a simple fire strategy developed to achieve compliance into one that really will provide a proper legacy. The knowledgeable fire strategist can make this be the case.

CHAPTER 4.
THE RULES OF ENGAGEMENT

"Imagination rules the world."
—Napoleon Bonaparte, French military and political leader

Even in war, it was recognised that there should be rules of engagement. This was formalized in 1864 in Geneva, Switzerland, when a group of world leaders and statesmen agreed to formulate a set of rules in order to reduce suffering and atrocities associated with military action. This became known as the Geneva Convention. Similarly, the business environment is required to operate under a certain set of rules; otherwise, unscrupulous habits would become even more frequent than they are now. It could be said that rules are there to provide the following functions:

- *To regulate the way we operate.*
- *To ensure consistency in our approach.*
- *To ensure best practice, whether technological, practical, ideological, or ethical, is appropriately incorporated in what we do.*
- *To ensure fair play.*

FIRE STRATEGIES - STRATEGIC THINKING

The international fire sector is fortunate to be blessed with a huge number of rules, codes, standards, and regulations covering practically every aspect of fire safety and protection. Some would say that there is almost too much in the way of guidance, as overlap and contradiction is not unknown.

From an international historical perspective and from personal experience, the majority of fire codes and standards used today derive from either the United States (NFPA) or the UK (BSI). (Note that these national codes may have, in turn, derived from sector-based sources, such as insurer's rules). Most countries have their own set of regulations, but they may be based on standards from other countries.

There is an additional layer of "sector" guidance and standardization. Airports, museums, hospitals, oil and gas processing plants, mining operations, and so on all tend to have their own specific requirements. Then are the requirements of stakeholders from insurers to building control to heritage bodies.

So, what does the fire strategist need to do to navigate around this maze?

Possibly the best way to start is to divide all possible rules into two tiers; those that are mandatory, and those that support or guide the fire strategy. I believe that a key input into any fire strategy is what I referred to earlier in this book as the mandatory framework. This is the framework that must be adopted by the fire strategist and will normally include national legislation, sector-based regulation, and any rules imposed by interested parties, such as the insurer.

The second tier would be those documents that support or guide the fire strategist. They may be there to provide detail and clarification to the mandatory framework, as is the case for many national standards, or may be there to assist with the preparation of the fire strategy but are not specifically required or called up. Again, national standards may also fulfil this role. The international fire sector "library" of advisory and guidance documentation is huge.

The quotation at the beginning of this chapter, from Napoleon Bonaparte, linked the terms "imagination" and "rule." I believe that the successful fire strategist will be one who uses his or her imagination, despite the constraints imposed by the written words.

Since my earliest days in the fire industry, I have been involved in standard-making, so I am fortunate in being able to see for myself the standard-making process, the people who are involved, and how requirements are made. My background has been influenced by writing sector-based rules, notably for the insurance industry, and national-based standards preparation via my involvement with British Standards Institution as a committee chairman, a member of a number of committees, and an author of British Standards. Let us take a look at the process of making a British Standard, as based on my experience.

It starts with a need: to standardize a piece of equipment or system, or a way of working. The former could be referred to as a specification and the latter, a code of (best) practice. This need may be brought to the standards-making committee by an external party, a member of the relevant technical committee, or the standards organization itself. Once the request is approved, work will start in one of two ways: a technical committee or working group will be set up specifically to prepare a draft document, or an external consultant will be paid to write the draft. Note that a typical "fire" committee would be made up of a number of interests, including trade associations, fire authorities and enforcers, building control, research organizations, insurers and other interested bodies and experts.

If a draft document is to be prepared internally, then, as you can imagine, a number of persons will have to share the work. They may attempt to do this within the committee structure or may form smaller working groups. Bearing in mind that such work often takes second stage to the "day job," a document drafted in this way could take years before it finally gets published. A slightly quicker method of producing standards is employing a consultant to prepare the first draft document. In this case, a brief is prepared, and a number of consultants believed to be capable will tender for the work. Once awarded the job, the consultant will regularly report progress to a specifically set up working group who would drive the work through.

Once a draft document has been completed, it will be reviewed in detail before being sent out for public comment. This in itself could take many months. When the comments are received, and they could be in the hundreds, the painful exercise of collation followed by distribution to committee members will start. In some cases, I have been involved in task groups set up to consider comments on just one section of the draft. In order to redraft sections based on the comments, many more months may go by before a document is ready for publication.

The final document will then go through a ratification and approval process. In some cases this is not the end, as errors are often identified at the last minute. Rather than stopping a publication in its tracks, a separate document highlighting amendments to the text may be published soon after the publication of the main standard.

In recent years, BSI introduced a third fast-track method of producing standards, referred to as Publicly Available Specifications (PASs). These documents are sponsored by the organization wishing to produce the PAS. They draft the document and, with the help of BSI project management, push it through a series of stages and have it vetted by a carefully selected panel of experts. In this way, it would be possible to produce a standard in around a year or so.

I have been involved in all three methods. I watched years go past in the preparation of the 2002 edition of British Standard 5839 Part 1 covering fire detection and alarm systems. I was fortunate to have won the contract to write a part of the British Standard 7974 series of documents supporting fire safety engineering, PD4—detection, activation, and control. My company, Kingfell, sponsored the production of BS Specification PAS 911: Fire Strategies in 2007.

National standards can be arduous to produce, but the level of complexity increases still further when trying to write European and International standards. For a start, there is the need to produce documents in more than one language, which can create interesting interpretations, as translations can often change the subtle meanings of sentences and paragraphs, compounding problems down the line. Then there are the historical national differences in the way in which fire protection is applied. Due to this, European and International specifications for equipment have been easier to get to publication than the equivalent codes of practice.

> *Never forget that standards are written by people with human fallibilities.*

At this point, fire strategists may be asking themselves how the methods adopted in producing standards affect them. I would like to devote the rest of this chapter to understanding some of the issues that come out of using standards, rules, codes, and so on.

Codes, Rules, Regulations, and Guidance Documents—what comes first?

Different countries treat standards in different ways, and how they are used and enforced can vary. Whereas, for example, one country may refer to a national standard as a "standard", another may call it a "code of practice" or "rule for…" Typically they all amount to the same thing, although some could be mandatory and some could be advisory, and others are perhaps not so clear-cut.

THE RULES OF ENGAGEMENT

One key aspect of standardization is how a standard should be treated. Let us start at the hard end of standardization—legislation. We are normally confronted with supporting documents, sometimes referred to as regulations. Typically, a regulation is a minimum standard that we are obliged to meet. As the name suggests, regulations are designed to regulate the way we do things. A quick search of the definition of regulation in the legal sense will tend to focus around the following: "A rule of order having the force of law, prescribed by a superior or competent authority, relating to the actions of those under the authority's control." Consequently, if a document is described as a regulation, it is normally a mandatory requirement or a first-tier document, as I referred to earlier. The term "rule" follows a similar connotation in that it must be obeyed. A rule may also be used to support legislation or to enforce requirements by other stakeholders.

We then move to a term that is commonplace in fire safety: "Code." Now, in some countries, a code is equivalent to rules; it is a mandatory requirement. In others, a code is advisory, a document of best practice. Some codes are actually described as "codes of practice." These documents would normally be formulated to give the user the benefit of the experience and expertise of others. As fire safety is a subject affected by changing ideas and technology, a code would be most appropriate. A code would also be more relevant for design and installation, allowing some flexibility. Guides and guidance documents are typically aligned with the flexibility of codes of (best) practice and normally contain best advice.

Specifications are normally used for a system arrangement or configuration for specific applications, or to set key criteria for manufactured equipment such as fire detectors and control equipment. Other documents could simply be referred to as "standards," which could be mandatory or advisory.

Another way of looking at it is that mandatory documents use the words "shall" or "must," whilst advisory documents use "should" or "may." However, phrases such as "where necessary" and "where appropriate" are not uncommon. Who determines whether a condition or section of a document is necessary or not? That is something that needs to be agreed on by all members of "the team," even though the fire strategist should have the knowledge or experience to lead to the right conclusion.

> **What are the types of fire standard in your country?**

When it comes to deciphering the most appropriate documentation to support a given fire strategy, the fire strategist needs to ascertain, at an early stage, what are the "Tier 1" requirements and what are the "Tier 2" supporting documents.

Today, we have two fundamental options for the development of a fire strategy: we can use prescription or go for a performance-based fire-engineered solution.

THE RULES OF ENGAGEMENT

There was a time when all standards were prescriptive; they told you what to do and how to do it. This made life very easy for both practitioners and those auditing their results.

The intention of prescription is that the fire strategy either conforms to the standards or does not. The advent of performance-based codes has added another way forward, although there are those who continue to support prescription over performance-based engineering. So we are left with two approaches, which in turn provides for a third way—a hybrid approach using a combination of performance-based and prescriptive guidance.

Sometimes, it is easier to demonstrate the concept of the two types of design basis with a diagram. Figure 2 shows two pie charts, one representing a prescriptive approach and the other, a performance-based approach. Prescriptive rules are designed to be "black-and white," that is, there is a right way and a wrong way. In an ideal world of prescription, there would just be black and white, making the lives of the fire strategist very simple (and possibly very boring). But we all know that no building, or use of a building, is likely to allow for every element of the standard to be met in full, unless the building is extremely standard in its construction, layout, and use, or the standards themselves are not so precise as to pin down the designer on every detail. There is also the possibility that the standard, or the building, has not been properly understood and the outcome is full compliance by default.

The Prescriptive Solution

The Performance Based Solution

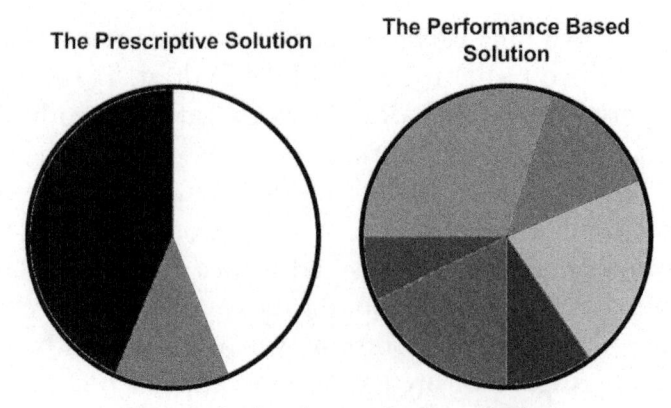

Figure 2: Analogy of prescriptive vs. performance-based solutions

If a prescriptive standard has been fully and thoroughly used, and the building has been properly assessed, there will be aspects that cannot be made to comply exactly, and are thus the "grey" area.

Sometimes, these grey areas are known as variations or deviations from the standard. Sometimes this is not seen in a positive way, and there may be a call to rectify those grey areas to convert them to full compliance. An alternative view is that the identification of non-conforming aspects demonstrates that the fire strategist has undertaken an appropriately thorough and objective exercise. The question then has to be, can we live with the variation, or should we provide for an alternative strategy to mitigate the additional risk introduced by the

> *Identified variations from a prescriptive standard— good or bad?*

variation? This is something the fire strategist will have to assess.

A performance-based approach will not lead to any obviously right or wrong answer. Instead, a solution may be more or less effective, that is, there will be degrees of grey with no absolute right or wrong approach. This is where the performance-based, fire-engineered solution could be seen as revolutionary in that it allows for flexibility and variety of options to be engineered by different people with different ideas.

This makes the subject of fire safety engineering exciting in that, by starting with a clean sheet, we can develop our own strategy in our own way. At the beginning of this book, I described the profession as "opinion engineering" and even used the term "black art." Now, as fire strategists, we do need to take one or more steps back to think about what our prime objective is. Are we trying to break fire safety design boundaries to provide creative solutions, or are we fundamentally required to assure ourselves that what we do will protect life, property, business, and the environment?

If we can be honest with ourselves for a moment, let us assume that we do come up with a fantastic performance-based fire strategy, one that appears to cover every aspect and provide for a solution that is inventive and has shown to be effective whilst saving cash when compared to more traditional solutions. If the fire strategy itself is well documented and follows proper reasoning, it is highly likely to be approved. Break open the champagne!

We all know that every electrical or mechanical engineering design gets tested during the commissioning and handover phases of a project and is subject to a period of observation after handover. Faults or misdesigns should quickly become noticeable. In the case of a fire safety engineered design, the only true test is a real fire, with two possible outcomes: a positive one where the strategy appears to have worked and a negative one where people have been put in danger and/or the building has been destroyed.

In the former case, the fire strategist could conceivably claim victory in that the strategy was shown to be effective. In the latter case, it would be extremely hard to pin the problem on the strategy. There are always cases where real fires just do not conform to the design scenarios. Nevertheless, performance-based approaches will improve with improving methods of analysis and with a structured approach to undertaking the assessments.

In summary, there are advantages in using the prescriptive approach:

- *Straightforward to use and apply.*
- *Based on past experience and research by others.*
- *Provides a consistent approach and output.*
- *Easier to assess and audit by other parties.*

And there are disadvantages:

- *Inflexible to the actual needs of the building in question.*
- *Will not necessarily lead to an optimum solution.*
- *May fall behind current design and assessment practices.*

Similarly, advantages of the performance-based approach include:

- *Can be tailored to the specific needs of the building, the processes and the occupancy.*
- *Allows for innovation, an ideal attraction for those with a need to push boundaries.*
- *Forces a holistic approach considering all aspects that may affect the building and its occupants. This book is all about a holistic fire strategy.*
- *Can lead to more cost-effective and appropriate designs.*

With disadvantages being:

- *Requires a more thorough and detailed assessment by suitably competent persons, which could initially result in higher fees and a longer front-end programme.*
- *The output may be more difficult for external parties to assess.*

And the final point sums up a potential dilemma: how can third parties be entirely sure that the performance-based fire strategy really does adequately cater for their concerns, or to put it another way, will it work? One answer to this is that the thoughts and ideas behind the approach are clearly and unambiguously explained so that it is not essential that every stakeholder has also to have high levels of competence in fire engineering.

In reality, many performance-based solutions could best be described as hybrid strategies; some aspects use a performance-based approach and others, prescriptive rules. Examples are where fire models have been used to set escape distance and evacuation models used to calculate escape times. The corresponding fire detection systems used to achieve these requirements are laid out in accordance with a prescriptive standard.

> *Have you ever been involved with a performance-based fire strategy without using any prescriptive standards?*

Whatever format or type of standard we use, we will, at one stage or another, come across prescriptive requirements. These could give maximum travel distances, fire detector spacing, minimum fire separations, and so on. Historically, standards have adopted a prescriptive approach to rule setting, that is, you will do this, this is the maximum accepted, this is the minimum accepted, and so on. But where do these rules come from? Can they be relied on?

I would like to give an example based on my experience of standards-making. British Standard Code of Practice 5839 Part 1 is used throughout the UK as well as many other parts of the world. It covers the design, installation, commissioning, and maintenance of fire detection and alarm systems. The current edition of the British Standard gives spacing requirements for the installation of fire detectors: any point in a room should be less than 7.5 metres from a smoke detector and less than 5.3 metres from a heat detector (in areas where smoke and/or heat detectors are required). So where did this requirement come from?

Well, many years ago in a British Standard Committee meeting, the decision was made to change the way we specified the allocation of smoke and heat detectors in a given area. Prior to the discussions, the rule was that there should be a smoke detector for every one hundred square metres of horizontal area, and a heat detector for every fifty square metres. Although this rule had been around for some time, it did not adequately convey the message that the concept is to ensure that a detector is never far away from a fire that could conceivably start in any part of a room. A simple area-to-radial conversion suggested that the appropriate monitoring radius should be 7.5 metres around a smoke detector and 5.3 metres for a heat detector, based solely on the existing area rule.

But where did the one hundred square metre and fifty square metre rules come from? Although I cannot claim the accuracy of the following statement, I was once told that when one of the detector manufacturers were awarded British Insurer's FOC approval for their smoke detectors, they were asked what should be the density factor for the devices. The response was that one thousand square feet per device appeared to work well. This figure was subsequently converted into metric—one hundred square metres. As heat is less transferable across a room than smoke, 50 per cent of the smoke detector area was chosen. It is as simple as that, so the next time an auditor fails a system because he or she managed to find a point in one room that was 7.57 metres from a smoke detector…

The same applies for many of the rules we take for granted. I understand that the thirty-metre search distance within a zone was originally based on the length of a manufactured hose reel.

Without knocking the way in which prescriptive rules were historically derived, most have stood the ravages of time and, furthermore, feel about right. When you observe smoke and heat detectors installed by the rules, it does not *feel* like they are too sparsely populated or too dense. Similarly, search and travel distances also feel right. Until we can prove that the measurement criteria are wrong, it is advisable that we keep with the figures.

A concession to the performance-based approach but definitely within the realms of prescription is the use of categories, enabling the expert to choose the most appropriate way forward for a given set of conditions. I would like to use BS 5839 Part 1 once again as an example.

Way back in time, the British Standard introduced a series of categories and subcategories for fire detection and alarm systems: Life Safety (L Type), Property Protection (P Type) and Manual Operation only (M Type). Both L and P Type systems were then subcategorized. The subcategories allowed for complete monitoring by fire detection in all parts of the building (L1 and P1), partial monitoring (L2 and P2), and monitoring of escape routes and the areas directly impinging on the escape routes for life safety purposes

(L3). Note that the life safety classifications were subsequently expanded to include L4 (corridor detection) and L5 (specialized fire safety engineered layout). The M classification indicated a system using just manual call points to initiate an alarm condition.

Let us first consider what we really need to determine the appropriate category. The first is the objective. Are we considering life safety, property protection, or both? This will come from deciding what the key objectives are for the fire strategy, as considered in chapter 3. In the case of this British Standard, I have seen specifications for L1/P1 by fire consultants who may not even appreciate what the key differences are between a life safety fire detection system and one for property protection. In this case, life safety systems are primarily designed to aid evacuation, which will include operating alarm systems.

Property protection systems are primarily designed to control the fire, which may include activating fire suppression systems and/or initiating a response from professional firefighters. Note that a pure property protection system need not operate alarm systems

> *"I always specify the highest levels of fire protection for my best clients!" claims John. What do you make of John's statement?*

around the building, only the alarms that will assist with the fire control operation.

Then there are sub-classifications, such as L1, L2, and L3. These were originally put in to allow for levels of monitoring commensurate with the building and the fire safety objectives for the building. Even though an L1 system requires complete monitoring of all areas, whereas an L2 is for partial monitoring, the former is not necessarily a *better* system than the latter.

This lack of understanding is fairly widespread, but it is vital that a fire strategist involved in making the decision understands the nuances behind categorization. The above example is taken from a British Standard, but there are instances all over the world. It is all to do with a proper understanding of what the standard is aiming to convey.

The last message I wish to make in this chapter is based on something that was indoctrinated into me from my earliest years in the fire industry, and that is to understand the spirit and letter of every type of standard. I believe the biggest cause of misinterpretation of any standard is that the standard is not properly understood by the person using it. It is the habit of taking clauses and phrases from a standard, in isolation, and applying them, without fully understanding the context in which they were drafted in the first place.

I believe that all fire strategists need to read and understand the standard they are applying, in full, at least once in their lifetime. Only in this way will they understand more than what the words say; they will understand the spirit of what is being conveyed. When I say the standard "in full," I really mean it. This should include the scope, foreword, introduction, and any other qualifying statements. In some cases, even the title of the standard is not properly understood. This is the only way we can make the best use of the "library" of rules, codes, regulations, and guidance available to the fire strategist.

In summary, the main points that I wish to highlight from this chapter are:

- *The fire strategist needs to prioritize and understand the mandatory framework from the support and guidance documentation. A two-tier approach may be a way of achieving this.*
- *Understand the nuances of the standards being used and the national connotations of those standards. Even if the standards are seen to constrain strategic thinking, if fire strategists have a reasoned argument, they can still be inventive in their approach.*
- *What is the design basis for the fire strategy: a prescriptive approach or a performance based solution? Don't forget the possibility of using both in one fire strategy.*
- *Understand the way standards are drafted and that there may be cases where the standard falls behind current philosophies and technologies. Understand both the letter and spirit of a standard.*

CHAPTER 5.
USING INTELLIGENCE

"All the business of war, and indeed all the business of life, is to endeavour to find out what you don't know from what you do."
—The Duke of Wellington, British Military Commander

Military strategies, or at least successful military strategies, make use of any relevant information that can assist in a successful outcome. Sometimes the information is easily available, sometimes it requires investigation, and sometimes it is misleading or incorrect. Such information is referred to as "intelligence," a word that is also used to describe the thinking abilities of the more successful living beings. Consequently, the words "intelligence" and "success" are often interlinked. The more successful businesses also make use of "intelligence" to gain competitive advantage and eventual profits, a key measure for commercial success.

When it comes to the preparation of fire strategies, I often wonder if the same level of homework is undertaken. I wonder whether the fire strategist chooses to delve into information that could support a better fire strategy. In fact, it could be the case that the only intelligence gathering is in the form of a "cut and paste" fire strategy—one that takes large chunks of text from a previous fire strategy and uses it for the current strategy. As the saying goes, "the clues are all out there," clues that can lead to a successful outcome.

The type and availability of intelligence will vary and will depend on the type of strategy being prepared. If the fire strategy is being developed for an existing building, then much of what is covered in this chapter will be relevant. For new build projects, similar levels of information can be derived from the analysis of the building users' fire safety management culture and/or the assessment of similar buildings and their use. Either way, this chapter may assist in the search for valuable clues.

A good place to start is with an evaluation of the existing fire safety management culture and, in particular, an investigation of the current or planned responsibilities and authorities for fire safety. This will give an instant clue to the fire strategist, illustrating how the eventual fire strategy will be managed. It can all start with an evaluation of how far up the organisational hierarchy persons with the responsibility for fire safety sit in the organisational chart.

Although those with ultimate responsibility are normally those at the top, it is not uncommon for those with the day-to day-responsibility for fire safety to be located far down the chain of command. In some cases, there may not even be the mention of fire safety, as this may come under the realm of "health and safety." In other cases, fire safety may be judged as a subset of an engineering or facilities management function. Then there may be cases where the function is completely outsourced or where nobody appears to have the responsibility for fire safety.

The organisational hierarchy can provide initial information as to whether the fire strategy will be successful. For example, if strong fire safety management is deemed a priority, then existing information needs to be found to support the idea that this will indeed be the case. If the fire strategy is highly reliant on active fire protection systems, then an organisation with a healthy budget for system maintenance will be a prerequisite. Where there is a need for the extensive use of fire compartmentation, then an organisation should have a good culture of snagging and signing off works to ensure that fire compartment walls and ceilings are not left breached by new services.

A good place to start is to review minutes of past fire safety meetings. Even the attendance sheet can provide clues as to how important the fire safety function is considered and can confirm whether the hierarchy is real or is an "ideal." There are several questions to consider: Who chairs the meeting? Which people regularly attend? Do you, as a fire strategist, believe that, from what you see and hear, fire safety is properly managed?

The minutes of meetings can reveal if any previous requirements or enforcements have been considered and implemented, and whether there are any substantial ongoing or planned actions to improve fire safety. The reasoning behind these actions could form a part of the eventual fire strategy. The minutes may also show aspects of the building and its occupancy that are not apparent and may save much time and resources in subsequent investigation.

There should be records, hopefully as part of a corporate report, of incidents due to, or affecting, fire safety. How these incidents are managed can offer some more clues that will be useful for focusing the fire strategy. Such incidents could include the results of minor or major fire incidents. How did the occupants of the building react? Did the management conduct the necessary follow-ups? This could also provide useful information as to the response time of the fire service and the performance of fire protection systems.

> *Is there a strong fire safety management culture? If not, this may affect the success of the fire strategy.*

A typical and commonly occurring incident around the world is that of "false alarms," a subject worthy of its own book. False alarms are a bane of modern fire safety and protection. So much so, that the vast majority of us typically assume that when we hear a fire alarm, it is a false alarm. Terms such as "unwanted alarm" have also been used to differentiate types of false alarm (basically, the fire detection system picks up fire-like signatures, but not from a fire that can cause danger). There is also guidance on acceptable levels of false alarms. What is highly relevant is how false alarms are recorded and treated. Many false alarms can be catered for by reviewing the type and location of sensors giving the false alarm, or the location and management behind the use of manual call points. If there are clues that actions have been taken following a false alarm incident, then this bodes well for the management of the fire strategy. If little has been done, then the fire strategy may be undermined by the management approach.

As well as the operation of fire protection systems, examination of fire prevention methodologies and procedures can provide useful insight into fire safety management practices. Some of these practices may be straightforward and documented and inspected. Others may just be an implicit part of the management culture. Corporate fire prevention policies are a good indicator of how fire safety is viewed. "No smoking" policies that were once used as part of the fire prevention strategy have now fortunately been supported in many countries by the legal implementation of "smoke-free" workspaces, primarily required for health reasons. Other policies can include the management and control of on-site works.

Considerations here may be the implementation of hot-works procedures and the control of site working by the requirement for safety risk assessments and method statements. As with smoking, legislation in many parts of the world has greatly assisted with support for other policies that were once optional, such as the procurement of fire-retardant fabrics and furnishings, fire risk inspections and audits, and fire safety training programmes. Nevertheless, the way these policies are implemented and recorded can provide useful insight.

Let us not forget that the fire strategy being prepared for an existing building, or for buildings similar to the one being designed, may not be the first fire strategy. Although the ideas behind a fire strategy may have changed, specific issues will remain the same, even if the building has changed fundamentally. It is possible that the fire strategy is not called a strategy but a plan, specification, manual, and so on. Any document that states the objectives for fire safety and how the necessary precautions and provisions are applied will provide useful intelligence and will save or reduce effort.

Many larger organizations may well have their own in-house rules and standards. As well as providing information supporting the fire safety management culture, the documents may well be a prerequisite for the fire strategy, part of the mandatory framework. Sometimes, multinational organizations may have generic international standards and local standards. The fire strategist needs to take both sets into account. It is at this point that any discrepancies are evaluated—not uncommon when trying to adopt international practices at a national level.

If the fire strategist believes that certain requirements are not relevant or are inappropriate for the fire strategy in question, these concerns should be raised at this stage of the intelligence gathering. Note that when questioning corporate standards, it is important that fire strategists have sound and objective

USING INTELLIGENCE

reasons for questioning them, not because they have a personal view that is at odds with their client's view.

In earlier chapters we considered the requirements of stakeholders such as the enforcer or the insurer. Let us not forget that their requirements are not always in the form of standards but may be incorporated in minutes of meetings, in correspondence, or in specifications. Once again, this could provide useful information, particularly when meeting with the stakeholders.

When we prepare fire strategies for existing buildings, the efficacy and function of current active fire protection systems should be assessed. Do the systems work as intended and will they support the fire strategy, particularly if the systems are a fundamental part of the strategy? It is easy to assume that the systems are not just operating correctly but that they are designed to meet the needs of the building. On many occasions, I have found this not to be the case.

> **How sure can you be that existing fire protection systems are fit for purpose?**

Clues can be found by reviewing fire protection system "operations and maintenance manuals" and maintenance records. These can help the fire strategist determine if there are any issues that can help with, or adversely affect, the eventual strategy. For example, specific and non-standard fire detection arrangements may have been the result of agreements made to please stakeholders. There may have been systems installed for property protection purposes, although this may not be explicitly highlighted anywhere.

We can start by checking that basic system components operate as intended—that fire detectors in alarm automatically lead to the operation of fire dampers, where required, or lead to the actuation of warning systems in the manner and areas they are required. This I refer to as a "fire system health check." Note that it is not just the operability of fire protection systems but the way that they are configured. It is not just that the fire detection system leads to the closing of fire dampers; it is more that the *appropriate* fire dampers are operated, and that the *appropriate* sounder systems are operated. In such cases, a detection system will cause certain events to be initiated with the effect that specific systems are triggered at the right time and in the right location(s).

The inputs and outputs can be shown on a table or matrix, such that specific inputs lead to specific events or outputs. A fire detector operated on Floor 10, for instance, could lead to the evacuation message being sounded on Floors 9 to 11 whilst all other floors receive an alert message. This is often referred to as a "cause and effect" matrix.

The "cause and effect" matrix is in fact the controller of the tactics used to support the fire strategy. The fire strategy is the document that sets the criteria, and the cause and effect matrix takes the criteria related to the fire protection part of the strategy and converts them into the operational model. Conversely, information related to the existing "cause and effect" matrix can provide invaluable intelligence to those preparing the fire strategy. If one assumes that the matrix has been previously approved by stakeholders, it can provide details of how evacuation zones have been set up, how protected routes have been allocated and protected, and how the interaction relates between, say, detection and suppression systems.

Cause and effect information should be obtainable from system handover documentation (such as operation and maintenance manuals) and from the fire system maintenance companies themselves. Note that this documentation can provide much relevant information and a feel for how the systems are set up and how they relate to the overall fire strategy. For new buildings, information from related or similar buildings can be of assistance.

The information is there—make use of it!

CHAPTER 6.
KNOW THE ENEMY

"In the practice of tolerance, one's enemy is the best teacher."
—Dalai Lama, Tibetan religious leader

Successful military strategists really understood their enemy. They knew their weaknesses and strengths and even how the enemy would behave under a set of circumstances. In just such a way, a successful fire strategy will take into account how a fire ignites, develops, and spreads in the context of the infrastructure of the building or site, referred to as the "physical terrain" in this book. The value of a fire strategist is in determining the most likely and/or most dangerous scenarios.

Fire development and spread has been scientifically studied around the world, and there are a number of rules that show the way a fire ignites and grows. Nevertheless, there is also a widely accepted understanding that a real fire will not always conform to these rules. This is not because the research and conclusions reached are invalid, but because a real fire can be affected by a near infinite number of parameters. We can never be absolutely certain how a fire will fully manifest itself, but we can predict how it will, most likely, develop. Fire behaviour is analogous to the behaviour of any living being, and this introduces an element of judgement, based on knowledge and expertise. Fire safety engineering is one of those subjects that cannot entirely be based on absolute science.

In many ways, prediction of how a fire develops and its impact on a building is similar to predicting the weather. As with fire, we know the key principles, and we understand the patterns of growth and movement. What we cannot predict with accuracy is what the end results will be and what will be the impact of a major event.

Many of us have come across the idea that a butterfly flapping its wings in one continent can cause a hurricane in another. Similarly, as a door opens on one side of a building, it is conceivable that a chain of events, deriving from the introduction of additional oxygen, may provide some stimuli to fire growing in another. Like the weather forecaster, the fire strategist cannot be absolutely sure of the impact but can use data gathered to create a best-case idea of what could happen. Once best-case scenarios have been chosen, the fire strategist's efforts will be to come up with a strategy to mitigate the effects of a fire, based on the objectives set.

> **Do fire strategists and weather forecasters have something in common?**

Experiencing a fire at close proximity brings home its real power. My first memory of feeling the effect of a fire was when attending a fire door test where a furnace test fire was applied to a fire door assembly. When the door finally gave way, and a gap developed in the order of a couple of centimetres, the impact was an immediate release of extreme heat—from ambient to "head in oven" temperatures in a couple of seconds. You suddenly get an appreciation of how "hot" a fire is. Experiencing a real fire at close quarters brings home the reality of what we are trying to control and can sometimes be a good reminder to those involved in fire engineering of what our subject is really all about.

The fundamentals of fire are simple. There are three elements required for it to exist: heat, fuel and oxygen. If any one element is removed, there will be no fire. Since in most cases, there is little we can do about the presence of oxygen, a fundamental rule for fire prevention is to control heat (ignition) sources, or limit fuel sources, or, best of all, both.

The way that a fire grows is predominantly to do with the way the three elements work together. Heat continuously applied to a fuel source will eventually lead to ignition. The point of ignition is dependent on the type of fuel sources. As the fire grows, the temperature in the enclosure of origin increases. This escalates further fire growth by increasing the ease of ignition of other fuel sources within this enclosure. Small fires become big fires, with the rate of growth normally dependent on the quantity and type of materials present in the enclosure of origin (the area in which the fire first ignited).

Fire manifests itself in three ways: heat, light, and sound. Heat is the primary energy source and is measured in units of joules. One of the tasks of a fire strategist, especially when using a performance-based approach, is to consider the rate at which heat is released. The rate chosen will affect the key outputs of the strategy, including evacuation and requirement of fire protection. The heat release rate (HRR) is measured in joules per second, or watts. To get a feel for the value, a small fire such as wastepaper basket fire will produce around three hundred thousand to five hundred thousand watts. Accordingly, figures of HRR are usually given in megawatts.

The determination of the appropriate HRR has created a type of subculture of fire dynamics analysis, with fire safety professionals arguing over the most appropriate figure for a given set of conditions. It starts with an assessment of the HRR of different types of materials. Test equipment known as the cone calorimeter is one currently accepted means to measure the heat release rate of materials.

The cone calorimeter was developed in the United States by the National Institute of Standards and Technology (NIST) in the 1980s. It uses the principal that oxygen consumption by a fire is proportional to the heat release rate of tested materials. As the name suggests, it consists of a cone-shaped funnel and heater that heats up the material specimen. The test equipment can also be used to assess smoke and gas production.

Although this testing does give us some useful information, it will only marginally assist with the assessment where a combination of materials is found together. This creates two issues for the fire strategist:

(i) *What HRR is the most appropriate for a given set of conditions?*
(ii) *What are the assumptions that need to be made about the quantity of materials in a given scenario?*

Sometimes there is a need to avoid opinion and guestimates. Where funds are available, it is probably relevant to undertake a test of an actual scenario to see what happens. Full-scale fire tests are used in specific circumstances and for certain types of risk profile. They can provide useful information that can support conclusions for actual fire development. Many aspects of a fire, from ignition to extinction, can be simulated for a complex set of conditions, although it has to be realised that it is a simulation and the parameters are chosen in advance.

The key benefit of full-scale fire tests is to allow for an appreciation of how a real fire reacts with a range of materials burning together, ordered in a way to simulate conditions of how they will be arranged in a real environment.

Despite the benefits of full-scale fire testing, there will always be a range of questions remaining unanswered. There will always be a need for the opinions of fire engineering experts. HRR is one area that has raised serious debate between experts. Particularly for performance-based fire engineering, the choice of HRR can have a long-reaching impact on subsequent decisions made.

I have spent many years in the rail industry and have been involved in a number of schemes covering surface and sub-surface railway stations. Typical scenarios chosen for investigation include fires starting within the confines of a railway station. We also consider the possibility of a train on fire entering the station infrastructure.

This requires an analysis of the peak HRR of a railway carriage. The way the carriage is built, its layout, and the materials used for furnishings will vary the value. The real test is that, in a fire scenario, a railway carriage may be full of the luggage and other possessions that evacuated passengers have left behind.

> *Choosing the right peak heat release rate is vital— trash in = trash out*

A fire strategy for a railway station will need to consider the scenario where it is possible that the carriage is within the station. The chosen peak HRR could greatly influence the requirements for fire detection and for smoke extract.

Given that smoke and toxic gases tend to increase with the increase in HRR, the value chosen can significantly affect the final fire strategy. Unrealistically low values would lead to an ineffective strategy, whilst higher values could lead to considerable over-engineering, with all the financial, technical, and logistical issues this will introduce. This is a typical conundrum that could be left unanswered, or could be subject to a best guess. As a fire strategist, it is important to derive an appropriate value, and check it with all stakeholders.

Whereas HRR is a key determinant in a performance-based fire strategy, the actual fire loading within a specific area such as a single fire compartment will need to be addressed for both performance-based and prescriptive approaches. The fire load is the amount of flammable material contained within the area under consideration, whether it is now and actual, or whether it is part of the design and use of the planned building and infrastructure. The fire load will determine the amount of heat and smoke generated.

A room's fire load is quantified as the amount of heat that would be generated per unit area in the room if all combustible materials present were burned. It is measured in kilojoules per square metre (or the imperial equivalent). An enclosure's fire load, in simplistic terms, is the quantity of the materials of each type divided by the area of the enclosure. The fire loading of an existing building subject to a fire strategy is one of the aspects considered by a fire risk assessment.

So far we have considered the inherent fire threat and the potential for a fire to grow. The next stage is to determine *how* a fire will grow.

Growth is normally depicted in graphical form as a curve from ignition to steady state. Fire growth curves will be found in many fire safety and fire engineering text books. The curves use the axes of HRR versus time, and general models follow a simplified "t squared" relationship. Given that different materials grow at different rates, a series of curves are normally used in fire safety codes, ranging from slow growth fires involving dense, hard-to-burn materials through medium, fast, and ultra-fast growth curves for highly flammable substances. When all the surface area of combustible materials is burning, a steady state of HRR is reached. This is followed by decay as the materials are consumed. We do, however, need to consider the possibility of flashover.

Flashover is a condition where heat radiating down from the ceiling smoke layer ignites not only the combustible material within the enclosure, but also the airborne gases. The whole enclosure will be subject to combustion and an instantaneous heat rise to a 1,000 degrees Celsius or more. The chances of flashover increase dramatically when oxygen is allowed to enter the mix, such as when a door is opened. The possibility of flashover could adversely affect the objectives of a fire strategy. The ideal solution is to not allow a fire to get to a level where the conditions for flashover exist. Strict management of combustible materials is one way to achieve this, but perhaps a more realistic way for the fire strategist is to ensure that the fire is detected early enough or can be suppressed within the enclosure of origin, before the conditions for flashover are there. Although fire codes approach the issue in different ways, ways of minimising the conditions and possibility for flashover should be considered in every case.

So far we have identified these key fire parameters:

- *What and where are the fire hazards within the building?*

- *What are the scenarios for fire growth? How fast will a fire grow?*
- *What is the likelihood for extreme conditions such as flashover?*

The growth of fire and smoke under a set of given conditions can be calculated from first principles. However, fire strategists need not get too caught up with the details, as there are tools out there to help them.

Computer-based modelling to assess fire growth and development around the building has, in my mind, come of age. A fire strategist can evaluate various scenarios using fire modelling programmes. The more sophisticated models will allow analysis of how smoke, and the various products of combustion, circulate around the building. I used the term "sophisticated" as there are free software packages available on the Internet that use rudimentary formula to show fire development within the enclosure of origin.

Zone models give an approximation of fire growth in relatively controlled conditions. They will largely be limited and may be appropriate for assessing pre-flashover conditions, or in some cases pre- and post-flashover.

Computational fluid dynamic (CFD) models can be much more powerful and can allow whole buildings or infrastructures to be assessed using numerous scenarios. Modern packages can allow analysis to show how, for instance, smoke extract systems can keep escape routes clear. Alternative designs, such as fire compartment layouts, can be separately evaluated and compared. Some can be linked to evacuation models to show the impact of a fire and the associated products of combustion on evacuating populations.

Ever-increasing computer power and more realistic programmes should bring the idea of fire modelling one step closer to mainstream use in fire engineering. I now believe that every fire strategy could make use of computer modelling.

What computer models cannot do is choose the input criteria and analyse the results. This is clearly something that the fire strategist will continue to do. Fire modelling will never replace the fire strategist, but it will be a fundamental tool to get more insight, thus assisting in the development of a more robust fire strategy.

Fire modelling will also not be appropriate for assessing the level of risk that the fire poses. A fire risk assessment is another input to the fire strategy that the fire strategist will need to oversee.

Where the fire strategy is part of the design process for a new building, the fire strategist will need to carry out an assessment of the building design, the proposed uses for the building, and the occupancy profile, together with an assessment of similar building designs and uses, to determine the risk profile.

For an existing building, a full fire risk assessment should be carried out to identify hazards and to recommend precautions to mitigate and protect against them. Where the objectives are purely life safety based, the risk assessment need only examine the risks for those who may be in the building at the time. Where the objectives also cover property, business, and/or environmental protection, the risk assessment will need to be suitably extended to evaluate these risks. The fire risk assessment should be a systematic and structured assessment of the fire risk in the building for the purpose of determining the current level of fire risk and the adequacy of existing fire precautions.

The subject of fire risk assessments is particularly topical in the UK, given that it is one of the main requirements of legislation. It has been found that the assessments on live buildings can vary considerably in content, scope, inputs, and outputs. Some may be extensions of an overall health and safety assessment, whilst others could be industry or even building specific to particular fire threats.

At the simplest level, a fire risk assessment could consider two independent variables, probability and impact, and assess how the combination of these variables leads to an overall risk factor. Probability will be the assessment of the likelihood of an event occurring in a specific room or area. Impact (or consequences) will be an assessment of how a fire in that area could affect the safety of the building and its occupants, or property, business, and environmental factors, where these are part of the objectives set.

Where a numerical or graded assessment is used, both probability and impact may be scored; the product of these scores provides a risk factor. Although a numerical system of risk assessment is useful for a comparative assessment of buildings, or of rooms and subdivisions within the building, it may not be useful in providing absolute guidance. In such cases, a qualitative assessment, where the risks are described, may be more appropriate.

FIRE STRATEGIES - STRATEGIC THINKING

There is often a difference between a fire risk assessment for the benefit of legislation and one to assist with the fire strategy. In a number of cases, the former could better be described as a fire compliance assessment where the assessor compares his or her findings with the requirements of building regulations and standards. The latter is likely to be more of a technical exercise where the risks could be quantified and determined by the team. The use of a risk assessment process in identifying both the likelihood and impact of a fire is sometimes referred to as probabilistic risk assessment. Note that probabilistic risk assessment may be much more complex than an assessment based on a subjective judgement of two factors and then multiplying them together. It can be developed using statistical techniques, which could be based on relevant fire data from historical fires. Assessments may use mathematical techniques such as regression analysis. We may also choose to apply sensitivity analysis to provide a more robust conclusion or use event tree assessments to show how a sequence can lead to certain outcomes.

> **Do you agree that a fire risk assessment is a required part of a fire strategy?**

Probabilistic risk assessments can be as straightforward or as complex as would be appropriate for a building, its occupancy, its process, and the objectives set for the fire strategy. There are alternative approaches to the use of probabilistic risk assessments.

Deterministic studies can be used to determine potential worst-case scenarios. To get to a point where this approach is useful, the fire strategist would need to derive a number of significant scenarios and ensure that the chosen fire strategy can cope with them. This can tend to be a risk-averse approach, unless good-quality data is used and fair and reasonable assumptions are made.

A comparative study of risk is another method used that, as the term suggests, compares the risk of the building subject to the fire strategy with similar building and occupancy criteria. This is a form of risk profiling that is a suitable method of fire risk analysis for new build projects.

Fire risk profiling can be a useful and quick way to categorize a building and its occupancy. The profile can be based on a single aspect of a building or may cover a number of factors, such as:

- *Building size/complexity*
- *Building use*
- *Occupancy profile, including numbers, age range, mobility, sleeping/no sleeping, risks, and so on.*
- *Average/worst-case potential for ignition.*
- *Average/worst-case fire loading criteria*
- *Typical fire growth curve*

Risk profiling tends to take a common sense approach in that, for example, one hospital would have a similar risk profile to another hospital, one petrochemical plant would tend to be similar in risk level to another, and so on. Data from one risk can help support a fire strategy for another, sometimes avoiding the need to go back to basics.

One key point that the fire strategist may need to get across to the stakeholders is that one risk profile may require more resources to mitigate the risk to an acceptable level than a lower risk profile. Similarly, the same level of resources applied to two different risk profiles will leave different levels of residual risk. In fact, I believe that risk profiles can be illustrated by curves, not dissimilar to the supply and demand curves that every person who has studied economics has come across. These are shown in Figure 3.

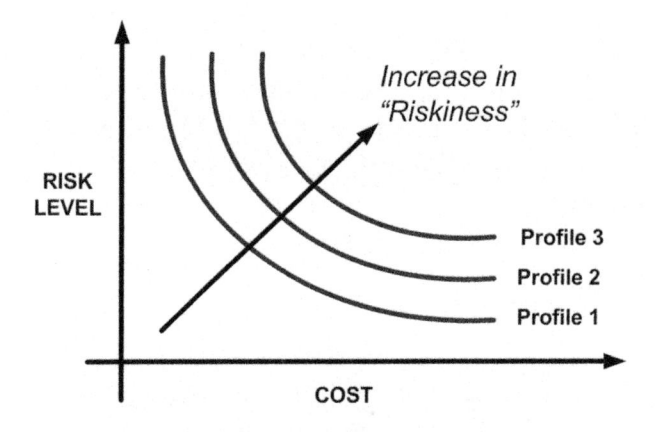

Figure 3: Fire risk profile curves

The coordinates are "level of risk" and "cost." Let us take one risk profile as an example. We should start with the idea that with minimal cost associated with applying fire safety and protection resources, there will be an understood level of risk. As resources are purchased and brought into the strategy, the level of risk will gradually be reduced until an optimum level of risk versus cost is reached. As more cost is invested into further risk mitigation provisions, the level of *additional* risk reduction will reduce until it reaches a stage where there will always be a residual level of risk.

Note that buildings of a certain type and use will inherently have different risk profiles, so the risk/cost curve will be on a different plane—the riskier the profile, the more the curve moves to the right. The purpose of the curve analogy is to promote an alternative way to think about risk, particularly when cost of the fire strategy is a measure. It may not be part of the formal strategy process but can be used to illustrate the concept of risk profiling to other stakeholders. There is a lot more to be found on the subject of risk profiling. Those who would be interested in seeing it applied should refer to British Standard 9999.

This chapter is all about the one thing we, as fire strategists, aim to control in order to meet our objectives. Fire does follow certain rules when analysed in its purest form. We have a multitude of research data showing how fire ignites and grows. But when exposed to a combination of conditions, as found in a living, working building, it can be difficult to predict. Fire modelling can, and will, help the fire strategist with this analysis, but there will need to be interpretation of the results. This is where the right decisions made about the fire parameters will influence the remainder of the strategy. The fire strategist will need to introduce a high degree of discipline and objectivity when determining what criteria to use:

FIRE STRATEGIES - STRATEGIC THINKING

- *Where will the fire start?*
- *How will it grow?*
- *What will its impact be?*

Fire strategists who rely heavily on an opinion or a subjective view, without substantiation, may find that their decisions do not stand up to subsequent scrutiny should the unthinkable occur.

CHAPTER 7.
THE PHYSICAL TERRAIN

"Understanding the six kinds of terrain is the highest responsibility of the general, and it is imperative to examine them."
—Sun Tzu, fifth century BC Chinese military strategist

A strategy, any form of strategy, will only work if the terrain has been properly examined. A military strategy utilises intelligence gathered from where the campaign will take place. All features of the terrain are closely examined in order to determine how the strategy will be most effective. Obstacles will need to be overcome, and key features should be used to best effect. Only when you properly understand the terrain and optimise and focus on each aspect should you then apply the necessary resources. A business strategy requires the same discipline. A business will only be successful when it understands the industry in which it operates. It understands the needs of its client and the actions of its competitors.

Take these same principles and apply them to a fire strategy. You can expect better results than if you simply took the features of the terrain as "givens" and did not make best use of what there is. This will require a more investigative style than the simple application of rules. For the purposes of this book, the terrain is divided into two parts:

- The physical terrain: This incorporates the building or collection of buildings, including their location and layout, their infrastructure, their structure, their inner subdivisions, their fabric, and their fixtures and fittings. These I will refer to as the static building elements. It also includes the processes. This is the dynamic function and use of the building(s). The assessment should include consideration of how they would contribute to both fire loading and ignition sources, and how they can affect evacuation and firefighting operations.
- The human terrain: This incorporates the various types of occupancy in the building and how it should be assessed. This is covered in the next chapter.

The building, plant or other built infrastructure will always be the primary focus of the fire strategy in that its design and layout will affect all four of the key objectives of life safety, property protection, business continuity, and protection of the environment. When we look at the building subject to the fire strategy, we need to understand its key features and how these features interact.

The location of the building and its context within the immediate community will be a good place for the fire strategist to start, whether it is a new building being assessed or an existing building. Starting with geographical or structural features, it should be ensured that these features will not impede any aspect of the strategy as it develops. A key consideration when it comes to location will be access for the emergency services, particularly where the strategy requires their early response. It is not unusual to find fire strategies that simply assume that firefighters can attend within a stated time period. Where emergency services are located remote from the building, issues as basic as road congestion could conceivably require a fire strategy rethink. Imagine you are preparing a fire strategy for a large shopping mall and that mall is extremely popular at certain times of the day, when roads leading to the mall are filled with vehicles attempting to park.

How would you account for this in your fire strategy? Possibly you may not see this as an issue. You may acknowledge this issue but not do something specific. You may decide, however, that this is worthy of a solution, and you will find a solution. If nothing is done, the fire strategist may need to revisit the attendance times and adjust for such a worst-case condition.

Another location issue is the potential impact of a fire starting in a neighbouring building. This is normally picked up in strategy formulation as many prescriptive and performance-based guides highlight the potential for fire spread (such as via radiative heat exposure) from a neighbouring building, or even from a potential explosion. There are the obvious clues, such as when you find that the neighbour manufactures fireworks, or the less obvious, where the building next door may be exposed to an increased

FIRE STRATEGIES - STRATEGIC THINKING

risk of arson attack. Fire strategists can determine key risk issues relatively quickly by asking the right questions as well as undertaking their own assessment.

Then there may be practical issues such as obstacles or obstructions that may impede the evacuation of persons away from the building and entry of emergency responders. Again, this is sometimes overlooked.

Getting persons to the exit point is normally the key focus, although it is also important to ensure that the route to the final designated place of safety or assembly point is also considered. Simple issues such as lighting external parts of the route could be used to great effect.

> **What locational issues do you find most common when assessing a building or plant?**

There is the potential for the ground around the building to become impenetrable due to adverse weather conditions. Even simple factors such as muddy paths or even flooded car parks can create real difficulties when looked at from a practical perspective. Again, this is a point that can be overlooked. If there are features that could prevent firefighting operations from some parts of the building, then the fire strategist will have to determine if this is an issue and determine how this can best be overcome.

We now move to the building structure and methods of construction. The subject of structural fire engineering is an area where specialist skill and knowledge are absolutely necessary, something often outside the realms of the generalist fire engineer or fire strategist. With the variety and complexity of modern building designs and with the increasing range of materials, the methods and philosophies behind structural fire engineering has needed to keep pace. Prescriptive and historical concepts may only be usable for those buildings that actually use traditional materials and structural methods. For many of the large buildings being built today and for those existing buildings that can be described as complex, performance-based structural fire engineering is probably most appropriate.

It starts with the choice of materials for the key structural members, which could be wood, concrete, steel, or a combination of materials. From there, an assessment is required for the way the vertical components, the structural floors, and the external facade influence the overall design criteria and building stability in a fire condition. This requires real skill and experience, as it is not just the way the key structural members independently act and react in a fire but how the structural elements perform collectively. As well as potential collapse scenarios caused by a fire, factors such as membrane action, which can pull and push the structure, will need to be appropriately assessed. We must not forget that even distortion of part of the structure can affect the ability of the building to maintain the key ingredients behind a fire strategy using active and passive fire protection. Consequently, basic objectives such as evacuation and firefighting could be impaired.

To support, enhance, or modify the calculus behind structural engineering, live fire tests can provide useful information.

Conducting full-scale tests of building structures can be a very costly and time-consuming exercise, and in many cases, you have only one chance to ensure the parameters are correct. The fire scenarios to be tested will have to be evaluated and agreed on by all parties, as the option of repeating some of the larger tests may not be commercially feasible. Even if the tests prove positive, it is then necessary to extrapolate the results to determine the performance of the full-sized building, which is a very delicate process that requires highly nonlinear analysis and critical thinking.

This once again requires specialist knowledge. The key objective is to ensure structural stability so that building occupants can evacuate from the building and to a place of safety. Consideration also needs to be given to allow time for firefighters to enter the building to either fight the fire or undertake search and rescue missions.

An assessment of structural stability and possible minimum time to failure under specific fire scenarios will need to be made, as this will affect the evacuation and firefighting requirements of the strategy. For new build projects, the building can be designed to cater for expected worst-case evacuation and firefighting scenarios. For existing buildings, the evacuation and firefighting planning will need to be prepared to meet with the worst-case scenario of building loss of integrity, unless extensive structural modification of the building is planned. We also need to consider if and where structures adjoin other buildings. It is not just the scenario of fire spreading from one building to the next that can be catered for by appropriate levels of fire compartmentation, but the structural impact of a fire in an adjoining building on the building being assessed. The fire strategist will always need to take this into account and either raise this as an issue outside the scope of the fire strategy, or take the time and trouble to analyse the potential impact and come up with a solution, if indeed there is a risk. What the fire strategist should not do is ignore the issue.

The fire strategist will also need to evaluate the external linings of the building, including roofs and other external constructions. Two main questions for assessment could be:

(i) *What is the impact of an external fire on the building?*
(ii) *How could a fire starting within the building affect other parts of the building via the external linings?*

Point (i) should have been covered by the location assessment, although possibly the linings used were not considered and the location assessment was simply geographical. In this case, the scenario will need to be revisited to ensure that the linings can cope with the external fire threat. For point (ii), the scenario of a fire exiting the building at one level and returning at other levels via the external walls is, unfortunately, one that many of us have become aware of, particularly in modern high-rise buildings.

Note that potential failures are not just to do with the choice of materials but also with the methods of construction. This is often seen to be outside the scope of the fire strategy.

We can afford some variation when it comes to the positioning of fire detectors or the pressure curve for fire fighting water systems. We cannot afford the same margin for error when it comes to structural fire engineering.

Once the building structure itself has been identified, the internal layout should be evaluated. When considering the layout of the building, particularly when looking at the means of escape, it is a good idea for the fire strategist to simply think back to the basics of what that is fundamentally required—what are the "no brainers"?

As an example, let us consider what is fundamental to any fire strategy—that the layout and the inner compartmentation fulfil the needs of the objectives set for the fire strategy. Simply speaking, for a life safety strategy, we need to ensure that the escape routes are suitable for all persons to safely evacuate, and also that the persons will be separated from the fire until they are safely away from the risk of fire. This simple principle can then be broken down into a number of sub-principles, such as:

(a) The dimensions and layout of escape routes should be suitable for the number and profile of occupants of the building. Sometimes when we use prescriptive guidance, we ignore the simple reasoning behind the criteria specified. All we need to do is ensure that the routes are *appropriate*. Perceptions of what is appropriate can differ between people involved at a local and a national level. Dimensions will include the width and height of escape corridors and the distance needed to be travelled.

(b) The building is likely to be subdivided into a number of fire compartments and separations, the purpose of these being to prevent the spread of fire through the building for a predetermined period of time—their fire resistance. The separating elements need to provide an adequate standard of both insulation, to avoid the unexposed side of the element becoming hot enough to ignite material in contact with it, and integrity, to avoid the formation of openings or cracks in the element that allow flame or hot gases to pass through it.

When considering these principles, some questions that the fire strategist considers could include:

- *How does the compartmentation arrangement deviate from conventional guidance?*
- *How does the compartmentation strategy conform to the existing/proposed evacuation and fire protection requirements?*
- *How do the fire compartments and separations match with the features of the building?*

Are there any accepted trade-offs between fire compartmentation and the use of fire protection (such as sprinkler systems)? Note that trade-offs between two forms of system may not always be acceptable. This is where the fire strategist will need to get guidance or approval at an early stage.

When looking at the designated routes for escape, we can come up with further guiding principles, given that the objective is to allow occupants of the building to evacuate from the building in the shortest possible time and in minimal danger.

Sometimes, guiding principles exist in the relevant national codes but it worthwhile for fire strategists to have memorized the principles they believe are appropriate. From that point forward, each of the principles can be quantified either via prescription or performance assessment. Here are some of these principles that could be applicable for any fire strategy, for any building type, and in any part of the world.

- Every occupant of the building should have two available routes of evacuation. Where this is not possible, additional fire safety and protection provisions may be required to protect the single route of escape. Dead-end corridor conditions should be minimized.
- Horizontal travel distances to an escape route should be minimised. Horizontal travel should, as far as possible, be on the same level.
- Vertical escape routes should be protected from the effects of a fire for a predetermined period. They should be capable of allowing the numbers of persons to evacuate in accordance with the evacuation strategy and should not, as

THE PHYSICAL TERRAIN

far as possible, be reliant on electrical or mechanical means, other than those systems specifically required by the fire strategy.

- All means of escape should lead directly outside of the building (other than where there are designated places of relative safety). No means of escape should lead back into a potential fire-affected area.
- Where persons are not able to directly evacuate from the building, they should be able to go to a place of relative safety from which they can be assisted in their escape.
- All means of escape should be appropriately illuminated with routes indicated appropriately.
- Special provisions may be required for disabled or mobility-impaired persons.
- Escape routes from any occupant position should, as far as possible, be kept free from obstructions.

> *What guiding principles, from experience, would you add to this list?*

From these guiding principles, we should then identify the potential for fire, smoke, and toxic products to move around the building.

Our prime consideration here is that nominated escape routes can be kept clear for the duration of the evacuation period and possibly for the subsequent firefighting effort. Ideally, there will be enough natural ventilation to cope with the calculated quantity and rate of smoke production, but often this will not be the case. Forced smoke extract systems will need to be used.

Smoke extract systems can require substantial plant and space within the building to allow smoke to be collected and taken to an appropriate release point. For new build projects, subject to an agreement at the relatively early stages of design, most types of smoke extract system can be accommodated. For existing buildings, this may not be straightforward. It is therefore the job of the fire strategist to evaluate how best smoke and the other products of combustion can be kept away from evacuees for the required period. Where this is not conceivable, it may be the case that certain evacuation routes just cannot be used.

As well as controlling the fire and smoke we know about, the fire strategist should also consider routes, such as voids, where it is possible that fire and smoke will be able to travel without restriction or detection, manually or automatically. In such cases a fire could become well developed by the time it has been first detected, potentially throwing out any of the assumptions made for the fire strategy. Heritage buildings are a good example of this. It may simply not be possible to identify all potential fire and smoke paths unless you are fortunate enough to have the original construction drawings. I have been involved in the analysis of buildings such as these, and the first thought may be a belt-and-braces approach, identifying and applying measures for every point where it is possible that fire and/or smoke could show itself. This could be an extremely costly exercise and is likely to fail, as all it takes is one unidentified path to override the fire strategy. It should also be remembered that many older buildings did not have the benefit of climate control and had to design in features that would allow the building to breath. By cutting off these features, it is possible that the building will suffer as a consequence.

At this point, we should move to consideration of the internal linings of the building as well as the internal processes.

The reaction to fire properties of all building linings, including walls, floor, and ceilings, whether for an existing building or planned for the new building, should be assessed by the fire strategist. The assessment should consider ease of ignition, rate of heat release, surface flame spread rate, smoke production rate, and total potential heat release.

Typically, the surface flame spread and heat release rate characteristics of the lining material should be of a higher rating in circulation spaces and escape routes than in other areas. Similarly, furnishings should be assessed for their fire properties. Let us also not forget that, in many cases, the contents of a building could have more influence on the size and growth rate of a fire than the fabric. Building contents should be assessed as part of the fire risk assessment and hazard analysis.

When looking at building plant and engineering services, the fire strategist will need to consider at least the following points:

- *Do they constitute a hazard? This will normally be part of the risk assessment and hazard analysis process.*
- *How susceptible are key services to a fire? In this case, we will need to consider how they are separated from other risk areas.*

The fire strategist should evaluate if the plant will contribute to fire, smoke, and toxic gas movement throughout the building. This could be the case with ventilation plants.

There may also be a need to access critical services in the event of a fire. The fire strategist will need to account for this also.

I like to refer to those hazards associated with the building linings, furnishings, and plant as "environmental hazards"—static hazards that are found in the environment of the room or area being assessed and cannot be easily moved.

So far we have considered the fixed elements for the building, but every building is designed to be used for something, and whatever the use is, this can be referred to as the internal processes.

All internal processes should be examined to determine how they will affect the overall building risk. Internal processes may include manufacturing, test and inspection, operational, and computer processing areas. Offices, shops, art galleries, and hospitals are all instances where processes take place, any of which

THE PHYSICAL TERRAIN

could potentially contribute to a fire event. I would even expand this further to areas which the general public use. The general public, if you think about it, are one big process, introducing a multitude of possible risks that could affect the fire strategy.

Given thought to this, and how the information can be best incorporated within a fire strategy, the hazards associated within an enclosure will exist in one of two ways: the local environment of the enclosure (covering the static elements as highlighted above) and the process (covering the dynamic environment—people, equipment, operations, and so on).

In 1996, I devised a simple two-by-two matrix, allowing a fire risk (or "hazard") assessor to mark up an inspection using the matrix, as shown in Figure 4, to allow quick and easy identification of each part of the building being assessed. The quadrants are as follows:

FIRE STRATEGIES - STRATEGIC THINKING

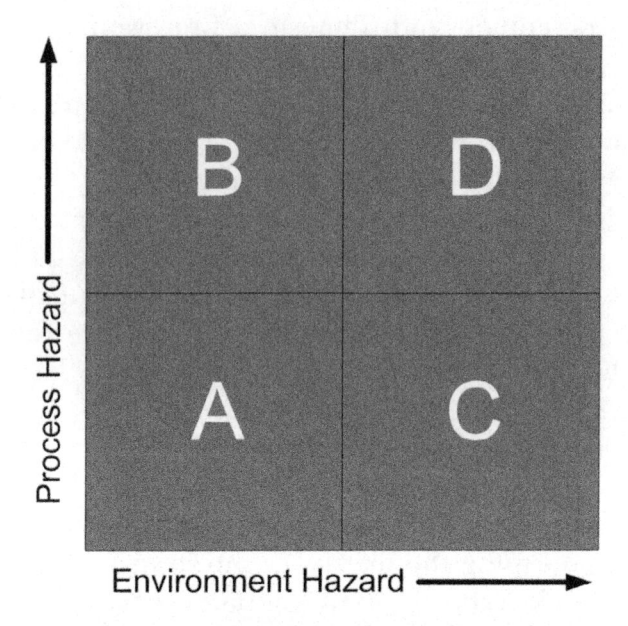

Figure 4: Significant hazard matrix

Quadrant A, Low Environment Hazard/Low Process Hazard: An ideal situation presenting little risk to the building.

Quadrant B, Low Environment Hazard/High Process Hazard: There will be areas where processes are necessary, so a low environmental hazard is appropriate.

Quadrant C, High Environment Hazard/Low Process Hazard: A good example is in the case of heritage build-

> **Look at the environment around you right now— where would you mark it on the hazard matrix?**

ings, where the inclusion of drapes, bookcases full of books, and so on presents a hazard, but the process within the area is controlled, possibly by good levels of fire safety management. Similar situations will be found in hotels, restaurants, and schools.

Quadrant D, High Environment Hazard/High Process Hazard: In such situations a fire is more likely to be an inevitability than a possibility. What often comes to mind is the Windsor Castle fire in England when a spotlight, used to undertake refurbishment works, set light to a drape. A simple "process" error led to a massive fire, largely due to the high level of static environmental hazards throughout the building.

By plotting each area, enclosure, or room on the matrix, a "hazard profile" can be prepared, rather like a hazard fingerprint for a building. In this way, the fire strategist can focus on key areas and devise a solution accordingly. Figure 5 gives an example of this.

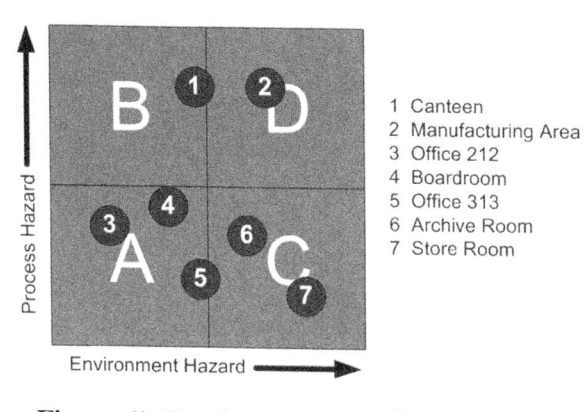

1 Canteen
2 Manufacturing Area
3 Office 212
4 Boardroom
5 Office 313
6 Archive Room
7 Store Room

Figure 5: Significant hazard matrix as used

FIRE STRATEGIES - STRATEGIC THINKING

Where each area of the building is marked onto the matrix is down to the judgement of the hazard assessor. Nevertheless, it is a comparative tool, so it should still focus the user on the areas where there is a potential problem. The fire strategist needs to understand the hazards, whether existing or on the design board. A fire strategy is incomplete without a proper appreciation of hazard.

Is fire hazard assessment the same as fire risk assessment? Not quite—when you assess hazards, you identify the areas where there is a definitive fire threat. With risk assessment, you assess the probability and impact of an actual fire incident. This was covered in chapter 6.

This chapter is all about the key focus of the strategy: the building or infrastructure. This subject, and all the considerations that go with it, could fill books on its own. There is plenty of prescriptive and performance guidance that covers every aspect of fire engineering and design for many building types. However, the focus should be on understanding what the key points are that the fire strategist needs to think about. Many of those points have been described here. The next chapter will cover what is arguably as important a feature as the building itself, and that is the people who inhabit it—the human terrain.

CHAPTER 8.
THE HUMAN TERRAIN

"Never underestimate the power of stupid people in large groups."
—George Carlin, comedian and observationist, United States

Even in the military use of the term "terrain," we normally think of the physical features of the place of battle affecting the success of the strategy. Military strategists have also had to consider the "human terrain," or the people that will be in and around the place of battle. This is particularly true where battles take place in urban or semi-urban areas. Sometimes the human terrain can provide surprises and certain "twists and turns."

When Adolf Hitler invaded Austria in 1938 to test his military strength and the appetite of Europe for occupation, he and his generals were pleasantly surprised by the reaction. Rather than being faced with bullets and grenades, they were welcomed with flowers. The human terrain in this case threw up a reaction that was not anticipated and introduced false promise

FIRE STRATEGIES - STRATEGIC THINKING

and a reinforcement of an invasion strategy that eventually, and fortunately, proved to be wrong.

It is likely that military strategists will evaluate the following parameters:

- *What are the numbers of people involved?*
- *Where are those people likely to be—will they be spread evenly over the area, or will they be concentrated in certain areas?*
- *What are their profiles (age, mobility, language, aptitude, and so on)?*
- *What is their likely reaction to events?*
- *What will be their speed of reaction and response?*

This is not dissimilar to the perspective of a fire strategy. One of the more important features of any fire strategy is to ensure that occupants of a building can escape safely in the event of fire. Even if the strategy were focused on property protection, it would be a major omission not to account for the safety of persons. This is true even if the building in question is a predominantly unmanned warehouse where contractors attend on occasion to maintain equipment. Their safety should still be paramount. It is a rare strategy that does not incorporate any element of life safety.

As in our exercise for the physical terrain, we should also consider some guiding principles for the human terrain:

Principle 1: All persons in the building need to be able to escape safely before the building collapses, or is at least no longer safe for occupation.

This requires us to get people from wherever they are, to a point where they are safe before the building becomes *unsafe*. We need to understand two parameters to determine this:

(i) *the length of time that the building is likely to withstand a fire before collapsing or becoming unsafe, and*

(ii) *the time required for all persons to safely evacuate from the building.*

Point (i) was considered in the last chapter and relies on the effort of both structural fire engineering and the other structural elements within the building. It also relies on what type of fire we are considering, where it is, and how big it is. Point (ii) is covered in this chapter and, once again, is based on a number of assumptions and conditions.

To meet this objective, we want the number associated with (i) to be much larger than that associated with (ii). But isn't this too simplistic an objective?

Even if the building remains standing, there is a good chance that everybody will not get out safety due to the direct and indirect impact of a fire within the building preventing escape. Furthermore, we may need the building to remain standing after everyone has escaped if firefighters are required to put the fire out. To cover these thoughts, let us introduce two more principles.

Principle 2: All persons in the building need to be able to escape safely before the conditions for escape become untenable.

FIRE STRATEGIES - STRATEGIC THINKING

This means that a fire is separated from those evacuating the building for at least as long as the required time for evacuation. This is normally achieved by the use of fire compartmentation, separating the risk areas from the evacuation routes for a predetermined time. We also know that it is not just the heat and flame of a fire that is the risk; the smoke and harmful products of combustion can also prevent a successful evacuation.

We therefore need to take steps to separate smoke and products of combustion from the evacuees whilst they are trying to escape. Part of this strategy is to ensure that the escape routes together with signage are visible; people do not walk freely through areas when they cannot see where they are going. They may at least become disorientated. This creates a requirement to keep the smoke layer above head height for the duration of the evacuation, and may be achieved with good fire separation supported by smoke containment and extract systems.

There are other ways of achieving this. The idea is to provide this strategy through to the point where evacuees are breathing fresh air, in a place where they are no longer exposed to the fire threat. Ensuring everybody can get to this ultimate place of safety may just not be possible. This could be influenced by the size and complexity of the building or the profile of some or all of the persons within the building. We need a fourth guiding principle.

Principle 3: The building structure is to withstand a fire to allow for firefighters to perform their function(s).

This may be to aid the evacuation of persons from the building (see Principle 4) or to control the fire so that it is no longer a threat. We need to understand the time that firefighters will need to get to the building and, once there, to undertake the necessary tasks. The combined time will need to be far less than the building collapse time.

> **What do you believe are the most important principles for life safety?**

Principle 4 (following on from Principle 2): If persons cannot be evacuated to a place of ultimate safety, they will need to be evacuated to a place of relative safety.

This is the point when an evacuation strategy may become less straightforward. An example of this arrangement is the use of refuges at designated points in the building. These refuges may be required, as an example, for the temporary holding of mobility-impaired persons. The evacuation strategy needs to ensure that these areas are separated from the effects of fire for the total duration of the time required to take them from this point to a place of ultimate safety. It must also be ensured that the route from the temporary point to the final point is protected and that those required to assist can do so also without the threat of fire impeding them from fulfilling their function.

Note that we could go on creating principles to cater for possible shortcomings or deviations introduced by the primary principles. By adding figures to the guiding principles, we have started to create a prescriptive standard. By adding performance requirements, we have started to create a performance-based fire engineering standard.

The exercise above is to prompt a way of thinking to allow fire strategists to go back to basics, to think for themselves rather than refer to standards without question. This is one of the points that I make throughout this book. A good fire strategist will be someone who thinks objectively, someone who can apply his or her experience, skill, and judgement to a set of circumstances where standards may not entirely fit. By going back to basics, we are forced to evaluate from first principles, and there is nowhere more appropriate than when considering the evacuation of people.

Evacuation analysis in the realms of prescriptive guidance was relatively straightforward. We knew the numbers of people in the building, or likely to be in the building, in the event of a fire. We knew the number and dimensions of the escape routes, both horizontal and vertical. We had basic criteria, such as the rate at which persons would use a staircase based on its width. We had figures for the average walking speed of a typical person. A simple calculation would provide us with an evacuation time. Easy to do, easy to check, and, to my mind, flawed.

One of the advancements of the performance-based approach is the introduction of two parameters: the available safe escape time (ASET) and the required safe escape time (RSET). The ASET can be determined by assessing the building, its processes, the way in which a fire will grow and spread, the performance of its passive and active fire protection, and so on. The RSET can be determined by assessing the flow of evacuees from the building. ASET looks at the impact of a fire on the building, and RSET looks at the impact of the building (in a fire condition) on its occupants. Needless to say, it is vital that the ASET be a far longer period of time than the RSET. What I love about the ASET/RSET model is that it promotes strategic thinking. The fire strategist needs to make a detailed assessment of both parameters.

The performance approach can help take away the "normality" of the traditional method of calculating escape times. My biggest concerns are the assumptions of "average" movement speeds and that persons will be equally spread over an area and that all escape routes in that area will be used equally.

All of us are different, and we all move at different rates. The elderly and young travel at a different rate from adults in their thirties and forties. A number of persons may be mobility-impaired and will have their own travel rate. Even taking ourselves, we have a variety of travel rates—whether we are late for an appointment, talking to a colleague on a cell phone, or under the influence of alcohol. There is no defined average speed as such. In fact, the idea of an average speed is entirely wrong. I put it to you that, in an evacuation, the average speed at which you travel is determined by others and not by yourself.

Let us take the example of an airport. We are all familiar with the travelator, designed to speed up and make easier the journey from one part of an airport to another. They are usually divided up into approximately one-hundred-metre sections so that we can jump on and off as required.

When we are confronted with using the travelator for the first time, we have the alternate option of using the standard walkway. We know that our walking speed together with the travelator speed will be the quicker option. Unfortunately, some fifty metres into our travelator journey, a group of fellow travellers have decided to stop and stand for the rest of the journey. We suddenly slow up and realise that the quicker option in hindsight would have been to take the standard walkway. At the next decision point, we may decide *not* to take the travelator. Although we had a good idea what our speed to the terminal gate might have been, our speed was affected by others on the journey.

This idea becomes more extreme at underground railway stations when we have no choice but to use a bank of escalators. Despite our wish to walk up the escalator rather than ride, our speed of travel will be dictated by the persons in front of us.

Both these cases are based on a standard non-emergency situation. A forced evacuation will not necessarily speed up events. Where large numbers of people find themselves part of an emergency evacuation, the situation could

> **How fast do you think you could travel through a crowded underground station during an evacuation?**

quickly lead to confusion, anxiety, and arguments—factors that could slow evacuation up even more. If we took the view that, within a group of people, the travel speed is dictated by the slowest person in that group, how would that change the calculation of evacuation times?

Another factor is the belief that we will use all escape routes equally in the event of fire. In fact, I do not think that anyone believes this, but it does make the simple calculation method easier. If we have two escape routes of three metres width each, and we have five hundred persons in that area, we could calculate the evacuation time by assuming that half the number of people will go to each of the routes. If only we behaved like that; you can almost imagine an emergency situation where someone decides to carry out a quick count of the occupants and put them into one of two evacuation teams!

The majority of fire strategists understand the commonly held view that we are all inclined to evacuate the same way we came in. This is simple human nature, as we tend to rely on what we already know first-hand, rather than what other people are telling us. Some fire strategies could take this behaviour and use it as part of the evacuation strategy. Others may try to use other means to force people to take other routes. This could be successful if the fire safety management strategy utilises trained personnel to guide occupants to the appropriate exit.

I've been involved with research projects introducing intelligence into signage systems to amend the typical initial human response in the use of evacuation routes.

Some of the initial research in this project showed that a relatively small percentage of us actually notice static fire signs.

What has always concerned me, though, is the subject of pre-movement time—that time interval between the alarm sounding and the commencement of the "planned" evacuation. When undertaking a performance-based fire safety engineered solution using ASET/RSET criteria, I often wonder if we really properly account for this period in time, which, in many cases, is a period of uncertainty. I have a number of thoughts about this that many fire strategists may be aware of or at least have concerns over.

When we first hear an alarm, what do we do? I do not mean what we are *supposed* to do, but what we *actually* do. From observation and discussion with other experts, our first reaction may be that the alarm is not really an emergency; possibly a test, or more likely a false alarm. I wonder why we react the way we do. Could it be that, in general, whether we are in an office or hotel, we have become used to being chucked out into the cold and rain whilst someone goes in to check "Zone 15" only to find out that it wasn't really a fire at all? Have we, as a community, become fairly sure that when we hear an alarm, it is most probably a false alarm?

I have pondered this long and hard. I have noted the exemplary efforts and research time spent by manufacturers of fire detection and alarm systems to combat false alarms. I have been involved with efforts by major users of systems to minimise false alarm rates. I have seen reports prepared by authorities highlighting how false alarms are costing the taxpayer dearly.

If we do get the false alarm rate to a really low level, can we then expect the general public to start trusting the alarms and react in the expected way? Something tells me that the answer will still be no. I have come to the conclusion that the reason why we tend to, or want to, ignore the initial alarm is more to do with the nuisance that the alarm means, together with a heavy dose of "it won't happen to me" syndrome. That being said, I still believe that false alarm rates can and should be brought down, as there are a number of factors outside of trust that will be improved by lower false alarm rates.

What I believe can improve the reaction to an alarm is the type of alarm that is used. I believe that there is a direct correlation between the reduction in pre-movement time and the closeness the alarm is to the human voice. A simple bell is likely to be confusing and possibly irrelevant to certain persons. A warbling sounder may be better to command notice, whereas a pre-recorded voice alarm message is likely to provide improved response. A public address announcement in real time is bound to initiate an action response that is much faster than the other methods, but someone shouting, "Get out of here—there's a fire!" into a crowded room will normally get the desired response. Possibly there is a relationship that could be subject to further research.

Another factor, based on observation, is that people tend to herd like sheep when they are not quite sure what to do next. Let us take the alarm situation again. A lecturer presents the subject of fire safety engineering to a class of, say, forty persons. The alarm goes off. What happens next?

FIRE STRATEGIES - STRATEGIC THINKING

If the lecturer calmly carries on talking about the benefits of fire detection systems, possibly we would face some inner conflict but we would also tend to be somewhat assured that the chief person in that room and at that time, the lecturer, must know what he is doing. We may be inclined to carry on listening. Perhaps after a while, with the alarm still sounding, the audience would start to reconsider their position. However, if the lecturer immediately stopped his presentation and made his way to the exit, we may similarly think that he obviously knows what he is doing and quickly follow behind.

What if the lecturer carried on with his presentation, and someone at the back of the room confidently stood up and left the room when the alarm sounded? The reaction time may be slightly longer but, in all likelihood, more persons would follow the confident person out of the room, and then the rest would follow.

The point here is that a useful aid to minimise pre-movement time is to understand the sheep and shepherd concept. Reaction times can be reduced by adapting a fire safety management plan to identify shepherds and possibly sheepdogs (those assisting the shepherds). We often refer to the sheepdogs as stewards, but, where there are no official fire safety stewards, alternative arrangements to lead persons will greatly assist the evacuation strategy. Whilst writing this book, I came across a facili-

> **How do you account for pre-movement time?**

ties manager who used the shepherd, sheepdog, and sheep concept to convey his own evacuation strategy. The shepherd concept will also be relevant where language and comprehension of alarm messages are an issue. It may be a factor when considering the practical arrangements for evacuating mobility-impaired persons.

There is another aspect when it comes to pre-movement time, and it is something that was highlighted earlier in this book. Sometimes, when we consider evacuation time, we look at the worst-case travel distance, say from a single dead-end condition, and then assume that this is the scenario appropriate for analysis. But what about those areas that are not directly on the means of escape? What about those areas high in a loft where regular maintenance is taking place? What about the basement plant or cable tunnels that need to be regularly inspected?

These places may not be areas that persons regularly inhabit, but the fire strategist really does need to consider the contractors, and how their evacuation strategy is impacted by working in confined or awkward-to-reach spaces. If these areas have not been considered, perhaps the pre-movement time could incorporate a factor for allowance for persons to get to the appropriate escape routes. It is a factor up to the judgement of the fire strategist.

When it comes to the actual evacuation sequence, there is a plethora of guidance, whether local, national, or international. I started this chapter with an appreciation of how these basic rules (principles) for evacuation can be formulated, whether a prescriptive or performance-based approach is used. It would come as no surprise to many that I am not a fan of the prescriptive approach. It is not that I am against the idea of prescription, more against the way it is sometimes applied, particularly when calculating evacuation times. In fact, when used well, and by those with an objective and analytical mind, the end results will be meaningful. Such a mind can derive answers that allow non-standard configurations to be assessed in a comparative manner.

My concerns derive from the use of prescriptive guidance more to prove a point than to provide a logical answer to a set of conditions. For instance, if the "required answer" is an evacuation time of, say, ten minutes, it would be possible to average out inconsistencies by making assumptions to fit the criteria. A bit of reworking and there is a good chance that the required answer is found. Sometimes the inflexible approach can be "broken down" to make it fit, whereas the performance-based method, properly followed, will provide a more realistic solution and will quickly show the "warts and all."

I had first-hand experience of this by assessing the same complex building using both methods. Using the prescriptive approach, I set up a set of basic calculations and proved that, lo and behold, the required escape time of six minutes was attainable, subject to a set of caveats. Previous assessments of the same building had produced similar results. I then used the same basic data and set up a 3D model of the evacuation routes using an evacuation modelling programme. We ran the programme and it showed, in real time, an escape time of eleven minutes. We ran the programme a number of times and could not get the period below this figure. Same number of people, same escape route dimensions. No doubt there are examples where the reverse is true, where the hand-calculated method produced higher evacuation times than the computer modelled method.

So, why such a major discrepancy?

To me the answer is very simple. As highlighted above, the hand calculation method can be tainted by opinion and caveat; the performance method relies upon academic discipline and a structured approach. This is not to say that the performance-based approach is failsafe. As with everything, the statement "trash in, trash out" applies totally. But I will go back to the term "discipline," and without discipline, anything goes. Then there is the small matter of the evacuation modelling software—can you trust it?

I suppose if the question were asked a decade or two ago, the answer would be different than today. Although I have not used it personally, my team have, and I firmly believe that it is the only way to properly assess evacuations, particularly for large or complex geometries. To me the key benefits are:

- *It will use the actual CAD (Computer aided design) layouts as provided. These will need to be built up to allow for both horizontal and vertical flow of persons. Once the layouts are built, it is then possible to adjust the layout, change dimensions of stairways, and add pinch points, the impact of which can be separately assessed.*

- *People can be profiled, a major and fundamental benefit of modelling. You can input a range of ages and levels of mobility, each of which will move at a certain speed and react in a certain way. It is the combination of large numbers of persons, all aiming to evacuate as a whole, that will allow for a more realistic result. Much better than "average" travel speeds. Adjusting numbers and profiles and the results obtained can be useful for identifying occupancy issues.*
- *Some software can allow for a more realistic scenario where certain escape routes are chosen above others—an analysis much closer to the truth.*
- *Some software can be combined with fire modelling programmes, allowing assessment of the combined impact of a growing fire on the ability and efficacy of those evacuating. Even aspects such as the impact on evacuees of noxious gases derived from a fire can be assessed in real time.*

There is bound to be a concern that such a fundamental part of a fire strategy needs to use an approach that can be appropriately validated. Many of the main programmes will have been validated by testing against live sequences and/or by peer review. When using any form of computer modelling, it is always appropriate to ascertain the type of validation carried out by the programme supplier. Note that the more robust modelling programmes will continue to develop as new research is found.

Let us put the actual evacuation effort to one side and ask a pretty fundamental question—evacuation to where? We all understand that the primary objective in an evacuation is to ensure that persons can get to a safe place away from any possible danger of the fire. If we state that the evacuation time is twenty-five minutes, this means that every occupant of the building would be safe from the fire within twenty-five minutes. Clearly this may not always be the case, so we need to actually understand where everybody is expected to be within this period of time.

A term often used, and which will be more appropriate as buildings become larger and more complex, is "place of relative safety." One of the purposes of the places of relative safety is to acknowledge that the evacuation strategy cannot conceivably allow for every person to safely evacuate the building to a place of final safety within the allowable time for evacuation, or in performance measurement terms, the available safe escape time (ASET). Such places serve at least two purposes:

- *Acknowledgement that some persons may need to evacuate to a place from which they can be taken to a place of final safety.*
- *The place of relative safety could actually be totally safe if the other parts of the fire strategy ensure that the fire threat can be eliminated prior to that place becoming unsafe.*

A place of "relative" safety obviously implies that the area may be safe for a while or under a certain set of conditions; consequently, the fire strategist will need to ensure that when specifying places of relative safety, this is closely tied into a management procedure to ensure that, at the right point and time, either those persons can move to an ultimate place of safety or that place can be shown to be safe when there is a fire elsewhere. Again, this may require an extremely robust fire safety management strategy to be successful.

> *Is the use of refuges becoming an essential part of a fire strategy for modern large or high-rise buildings?*

Refuges are a form of place of relative safety. These are locations around the building to temporary house persons with mobility impairment, or persons unable to go directly to final safety within the primary evacuation strategy. Note that refuges could be serviced either vertically, such as with stairs or lifts, as appropriate, or horizontally, which may involve persons traversing a floor of the building, possibly in a fire. It stands to reason that refuges, as with any form of place of relative safety, will be fire-separated from the rest of the building and will need to be intricately tied into the fire safety management strategy. Fire strategists identifying and specifying refuges without considering how they will be used have not fulfilled their role adequately.

This chapter has identified some of the issues of occupancy of the building as part of the fire strategy. As with the physical terrain, this subject alone could fill many books, and this is an area where research continues to improve our knowledge of how we react when faced with a fire. The key message here is that handling people in the right way is as hard as handling the fire itself. Research continues!

CHAPTER 9.
STRATEGIC VISION

"Strategic planning is worthless—unless there is first a strategic vision."
—John Naisbitt, author and futurist, United States

Every strategy should be based on a strategic vision. The vision is an idea, a belief, of how the strategy will deliver the objectives set. Military strategists would first have a strategic vision of how they see their campaign developing and what success will look like when they reach their objectives. Similarly, all successful businessmen have a clear vision of where they want to be. They will then develop their strategy to deliver that vision.

Strategic vision is all about turning an idea into something more tangible—a simple way of picturing what the strategy is about, and how it should deliver the solution.

Fire strategists may rarely see their role as a visionary, although, in reality, experienced fire strategists may have a good idea of how they see the fundamentals of a strategy progressing, based on what they know about the building and what their inclination tells them.

Note that I am not a fan of those preconceived ideas that some fire engineers have, when they move from one project to the next, a form of "cut and paste" thinking. What I do appreciate, however, is that wisdom that develops in all of us over time. Utilising that wisdom at the front end of analysis will save the need to go back to first principles. This wisdom is another way of describing strategic vision.

One of the ideas specifically developed for BS PAS 911 was a method of allowing a quick and easy way of "picturing" a fire strategy. Ideally the picture should be apparent using one side of A4 paper. This method is captured in a diagram (Figure 6) and has been designed to identify the main elements of a fire strategy. It is intended to allow the user of the diagram to identify how they believe each element contributes to the overall fire strategy. The original title of the diagram as given in PAS 911 was the "Strategy Value Grid," as it allowed identification of the relative value of each of eight elements appropriate for every fire strategy. As well as allowing the visualization of a fire strategy, it also allows for value analysis as the strategy develops.

STRATEGIC VISION

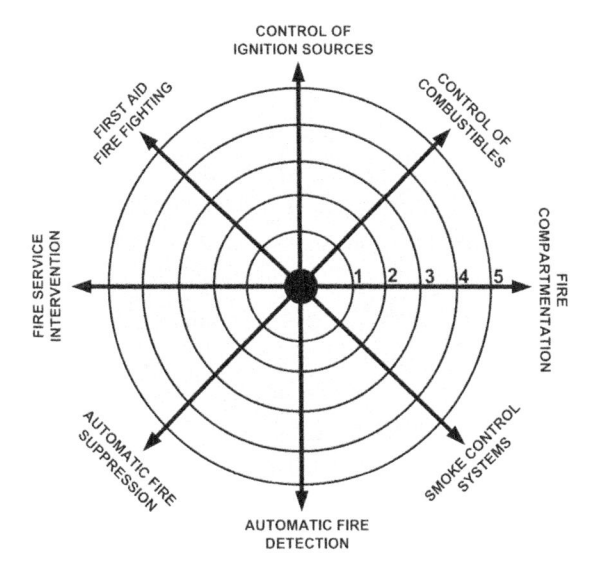

Figure 6: Fire strategy value grid

It all starts with eight key strategic factors:

1. Control of ignition sources
2. Control of combustibles
3. Fire compartmentation
4. Smoke control systems
5. Automatic fire detection
6. Automatic fire suppression
7. Fire service intervention
8. First aid firefighting

Note that these are *primary* strategic factors. Secondary factors such as alarm sounder systems, fire doors, extract systems, and fire prevention policies will follow from the primary factors.

The idea of the diagram is to allow each of the eight factors to be separately considered and scored from zero to five, based on their relative importance to the strategy. The diagram was designed to be used as a first round of analysis, although its real benefit is in regular revisiting it as the strategy preparation progresses.

It will allow the team, or some of the members of the team, to sit around the table and consider each of the eight elements and, based on their judgement, to mark each of the elements against the number with an "X" or dot. At the end of the assessment, the Xs are joined together to form a pattern. This pattern helps describe the type of fire strategy. See later in this chapter for that. I've sometimes described the diagram as "a fire strategy on one side of paper." What determines where the scoring goes is in effect based on the "terrain," both physical and human, as described in the earlier chapters. The following few pages describe each of the eight elements and how the terrain can affect the score.

Control of Ignition Sources

Ignition sources may be found throughout the building. They could be a fundamental part of those processes within or around the building, or may be brought in by occupants or any other groups of persons who use the building. What the fire strategist needs to understand is the possible and probable main sources of ignition and how they could combine with potential fuel sources. We need to undertake a paper-based or site-based "ignition" risk assessment, taking into account each constituent part of the terrain. Typical questions here should include:

- *What rules or procedures are, or will be, in place to control ignition sources?*
- *Where are the key "static" areas of the building where ignition sources may exist, and are they catered for by rules and procedures?*
- *How will specific fixtures and fittings contribute to potential ignition?*
- *For each process within the building, where are the potential ignition sources?*
- *Are any of the occupancy profiles such that they could introduce ignition sources?*

Ultimate control of ignition sources relies on proper fire safety management. Without this, this factor should be scored lowly and only increase where the fire safety management culture is deemed to be robust enough. This alone does not determine the score, as there may well be permanent ignition sources as part of the fundamental processes of the building, even if the fire safety management controls are first-rate. The fire strategist will need to determine the balance between fire safety management culture and the residual ignition risk.

Control of Combustibles

Combustible material, the fuel sources for fire, will predominantly exist in fixtures and fittings and possibly within the processes of the building. Furthermore, it is quite possible that fire loading will be introduced by some, or all, of the occupancy profiles. Relevant thoughts here should include:

- *What rules or procedures are, or will be, in place to control combustible materials?*
- *Where are the key "static" areas of the building where there are combustible materials that cannot be removed or completely controlled? Examples here include drapes in historic buildings.*
- *For each process within the building, identify where the fire loading is and its proximity to potential ignition sources.*
- *Are any of the occupancy profiles such that they could introduce combustible materials without fire safety management controls?*

As with control of ignition sources, control of combustibles relies on proper fire safety management, and, as with ignition sources, the scoring will largely be dependent on the fire safety management process. There are some buildings that are full of ignitable materials. There are ways to overcome this by replacing the materials with less flammable versions. In many cases, this may not be acceptable to stakeholders. Based on my experience with a number of fire strategies for historic buildings, museums, and art galleries, certain stakeholders often have the power to overrule even the most obvious suggestions for fire safety purposes. In such cases, unless the fire safety management is first class, and the statistics for this and similar buildings show a low fire risk, this factor cannot be scored highly. The fire strategist will need to determine if indeed it is possible to ultimately control combustible materials.

Fire Compartmentation

Fire compartmentation, or perhaps as more generically known, passive fire protection, was probably the original form of fire protection. The realisation that a solid brick or stone wall will segregate one part of a building from another in the event of a fire continues to this day. When it comes to more modern building design, the use of such solid building methods has declined, as the tendency for large, open plan areas, sometimes forming a multilevel atrium, means that passive fire protection on its own will no longer be appropriate. The two key reasons for compartmenting one area from another are:

- *To contain or control a fire for long enough to allow evacuation of persons from the building.*
- *To contain or control a fire for long enough to allow for time to suppress or extinguish the fire by automatic and/or manual means.*

There is a third idea that a sound and solid fire compartment will contain the fire until it is starved of oxygen, although this is not normally part of a fire strategy.

Sometimes the simple reasoning behind the need for fire compartmentation gets lost, and I have seen many elaborate fire compartmentation schemes where it seems that no strategy has been applied and that, where there is a suitably solid wall, it is deemed to be a fire separation line, with all the maintenance that entails. Consequently, plans may be shown with "red or blue lined" fire separations which are just not necessary as part of the strategy.

It must always be remembered that proper fire separations and compartments require constant maintenance. Walls, floors, ceilings, doors, glazing, dampers, and fire penetrations all require maintenance to be effective, often at some considerable cost to the building owner. Elaborate fire compartmentation schemes are potentially a real problem for the industry. Better to have a series of simple compartment lines fulfilling the needs of the strategy, where it can be ensured that they can be properly maintained.

Then there is the chosen level of fire resistance. Again, this would need to be linked to the evacuation strategy as well as the property protection strategy (where deemed appropriate) rather than an arbitrary number chosen simply because it can be achieved. It should be obvious that if there is a requirement for a forty-five minute evacuation, then the fire compartmentation protecting that route will need to maintain integrity for at least this period, together with a contingency factor. There is also the misconception that a higher fire resistance will protect against bigger fires. If there is a sound compartment line, it will protect the route for at least that period, independent of whether it is a small fire or a larger one. There is a difference between insulation, ensuring the temperature of a fire on the exposed side does not radiate or convect through to the unexposed side for the required period, and integrity, where the separating line will not buckle or break due to the effects of the fire for the required period. When considering the relevant construction materials for creating fire separations, fires that could lead to possible explosion will not be contained by a number of layers of plasterboard; it is better in such cases to use concrete or brick.

To summarise, where there are specified escape routes to a means of safety, these should be protected by fire separations in excess of the evacuation time of all persons. Where there are areas required to be separated from other areas for the purposes of containing fire for fire suppression, they will need to withstand a fire for at least the required period of extinguishment for automatic suppression systems and/or manual firefighting. Note here that many fire strategies may use automatic fire suppression as a knock down and still require manual intervention.

> **Fire compartmentation will only be effective if it can be properly maintained.**

One other point that can often be overlooked is that fire compartmentation needs to be thought about more on three dimensions than simply two. Elements such as constructional "dog legs" and differences in compartment lines between floors can mean that what appears to be a solid and consistent fire compartment line actually isn't.

Aspects such as this will diminish the ability to score highly. Other considerations are:

- *For new designs, does the need for fire compartmentation complement the intended design and use of the building? A conflict in specification could mean that the consistency of compartmentation throughout the building may be compromised.*
- *For existing buildings, is there a coherent use of compartmentation? Are there areas where the compartment line appears to vary from what is deemed necessary?*
- *Can compartment lines be guaranteed given that fire compartmentation is as good as its weakest point? Is there potential for voids that cannot be identified, let alone fire separated? Can it be assured that compartments are soffit to soffit and not just purely what you can see?*
- *Can it be ensured that the fire compartmentation can be properly maintained into the future? This will require incorporation into fire safety management practices.*

As an exercise, it may be useful for fire strategists to take a past project, possibly something they have not been involved in, and review the compartment lines as if they were there to meet common sense objectives. Is there a pattern? Lessons learned can then be used for their present project.

Although fire compartmentation may be a deemed key factor in the strategy yet cannot be guaranteed as per the issues given above, it cannot be scored highly, or the scope of the fire compartmentation boundaries is reduced to a level where assurance in its performance is higher, then the score can be raised up again. At this point, the fire strategist must be realistic and not idealistic; if it cannot be scored highly, then other means to ensure the objectives set need to be considered.

Smoke Control

As with fire compartmentation, smoke control aims to provide for similar objectives: to protect the escape routes and to allow for firefighting (in this case, manual firefighting). The thing about smoke control is that it *has to work*. A smoke control system that doesn't quite meet its objectives will be one that is next to useless. We could start with the very simplest methods of controlling smoke, such as opening a window or vent. This may provide some relief, but in most cases, and with the majority of modern buildings, this will not be enough. The essence of smoke control systems for life safety purposes is to keep escape routes free from smoke. This does not mean that all smoke will need to be extracted for the required period but just enough to ensure that there is visibility to allow escape, that is, the smoke and the associated other products of combustion should be above head height. This means that for automatic smoke extract, we will need to know with sufficient accuracy how much smoke is likely to be produced.

As we highlighted in chapter 6, in a real-life fire, we cannot be absolutely sure of the fire size. Consequently, we should base designs on a worst-case fire size. Only in this way can we have relative certainty that the system will work. Furthermore, automatic smoke extract systems work best when the smoke is hot so that it rises vertically and can be extracted at a high level. Smoke that quickly dilutes into the surrounding air will not have such buoyancy, and thus it may be more difficult to get the above head height smoke layers that we want.

Most of this is well known to many fire strategists, but how smoke extract systems are used can be a cause for concern. Given that smoke extract should allow for additional time for escape, it is conceivable that the systems could be used to allow for reductions in other factors, such as extended travel distances and removing "unnecessary" escape routes. The point here for fire strategists is to be sure in their mind that automatic smoke control systems will be effective. They must decide if they will indeed provide for extract for the fire sizes possible. The dimensions and layout really do lend themselves to the installation of a system.

Smoke control should only be scored highly if there is a defined need for it and the building can accommodate the required designs. Also to be taken into account here is the maintainability of the system(s).

Fire Detection

As with fire compartmentation, automatic fire detection is often considered a cornerstone of any fire strategy. Although the first reaction to score highly will be a temptation (after all, how can any fire strategy *not* have automatic fire detection?), it would be a good idea to consider why we need fire detection, and what it is actually there for.

The first premise is that fire detection systems, on their own, do not actually do anything. In fact, fire detection systems are not a subset of fire protection systems at all; they are simply fire-monitoring systems. It is what you do with the signal once a fire has been detected that is relevant—to initiate warning systems and/or control fire protection systems. However sophisticated the fire detection system may be, and whatever type of fire detection system is used, they are effectively there to replace the human senses. If we could sit a "reliable and awake" person down in every room and corridor in the building to be monitored, then there would be no need for a fire detection system. If I had written this twenty years ago, I may have also concluded that people are more reliable fire detectors than systems, in that they have the capability of picking up a fire that is potentially dangerous as opposed to one that isn't. They would be able to discriminate between fire and other non-fire or unwanted phenomena. Today I am not so sure. The sophisticated devices and systems are probably up there with the human senses, and they are unlikely to fall asleep, if properly maintained.

The question therefore goes back to this: do we need a complete system throughout the building telling us if and when a fire has been sensed?

The idea of the eight-pronged spider diagram is that if the strategy is deficient on some aspects, these could be made up by reinforcement of other aspects. To put this into context, if controls on both combustibles and ignition sources are very high—both scoring five—what is there that needs detecting? If we have managed to absolutely control sources of heat and sources of fuel, then there will be no fires, so, ipso facto, there will be no need for fire detection systems, or for any form of fire monitoring or fire protection, come to that.

A strategy that does not include the need for automatic fire detection is not in the realms of fiction. Only in the last few months, I have come across a very effective fire strategy relying totally on a first-rate fire safety management philosophy. There was no automatic fire detection system and not even manual call/fire points. Even after asking a number of "what if" questions, the strategy appeared to be foolproof. I would put my name against such a strategy. When considering the importance of automatic fire detection for the strategy, some thoughts may include:

- *Is there a defined need for automatic fire detection based on the requirements for the strategy? Is there an alternative method for raising warnings or controlling equipment?*
- *Is there a requirement for complete monitoring in every part of the building, or would partial monitoring achieve the objectives of the strategy, given the status of other factors of the strategy?*

- *Are there any constraints on the siting, installation, commissioning, and future maintenance of fire detection systems?*

Fire Suppression Systems

There is a vast array of methods to suppress fires, and more ideas and techniques are being developed all the time. Note that the term "fire suppression" is used rather than "fire extinguishing," which was the way such systems were described when I first joined the industry in the 1980s. This is based on the idea that the system may primarily be used to control and keep the fire in a controlled state until it can be fully extinguished or until it can be *confirmed* that it has been fully extinguished.

This does not preclude the idea that fire suppression systems can, and do, extinguish fires to the point that there will be no re-ignition. If fire suppression is a key part of the strategy, it should be determined whether the system(s) will be designed to fully extinguish a fire, or to control a fire until it can be confirmed that the fire is dead. This confirmation may be required from professional firefighters. If this is the case, the strategy will need to ensure that both aspects have been allowed for.

Fire suppression systems will need to be appropriate for the likely fire loading. In the vast majority of buildings, an automatic fire sprinkler system, when properly designed, will always be effective. It often surprises me that, well over a century after the introduction of such systems, they are not more widely specified. Fire sprinkler systems have shown, time and time again, that they can control and extinguish a fire at source, prevent fire spread, and reduce the threat from toxic gases and from flashover. Typically their benefits as part of a package of property protection are known, and the international insurance industry has been behind the rules and regulations covering sprinkler systems for most of the last century and into this century. More recently, the life safety credentials of automatic sprinkler systems have been recognised, and the use of systems as a trade-off to allow for greater fire compartments or longer travel distances has been included where fire engineered design solutions are used.

Today's choice of fire suppression extends from a number of water-based options including fog/mist systems, deluge systems, and other specialist systems. There is a range of gas extinguishing systems, proving that the demise of Halons has been good for the industry. There are air volume inerting systems, powder-based systems, and systems that react with the free radicals of flame. A large choice of systems exists, and it is getting larger, although there sometimes is a problem that I have personally encountered getting some good ideas to market. Sometimes a niche product, however effective it may be, may never get to market because there is no scheme to test it to, and there is no market to justify a scheme in the first place. Testing a new form of fire suppression is costly and time-consuming, and this can be a real problem if fire strategists would like to increase their armoury of resources.

But the question fire strategists need to ask themselves is whether there is a need to incorporate automatic fire suppression systems. If there is a property protection, business continuity, or even environmental protection objective, then it is highly likely that such systems should be employed either partially or totally within the protected building. Even for life safety purposes, automatic fire suppression systems may play an important part, even if it allows for reduced levels of fire compartmentation.

Fire Service Intervention

Sometimes we all assume that the intervention from professional firefighting services is a given. It is often assumed that the fire brigade will be in attendance in "X" minutes, as that is the guidance given to us. Well, one thing that I have learned along the way is that assumption is the mother of all mistakes!

If the fire strategist determines that this should score highly in the strategy, then there are a number of considerations that need to go with this. For a start, confirm what the expected attendance time is and check if there are any conditions to this (time of day, for ex-

> *Emergency service attendance times—never assume, find out!*

ample). Then ensure that once the emergency services are on their way, the process from arrival to fighting the fire is as smooth as possible. Consider where the firefighters will arrive; how will they be met? How will they access the building, including upper and basement areas?

Don't forget that if this is a key part of the strategy, it does need to be ensured that the firefighting infrastructure is available for the firefighters to utilise. In some cases where professional levels of emergency response are essential, and the size and complexity of the building(s) warrants it, it may be appropriate to set up an in-house professional firefighting team.

First Aid Firefighting

Where do we *not* come across portable fire extinguishers or hose reels or fire blankets in buildings? Legislation and guidance tend to mandate that first aid firefighting equipment be provided practically everywhere outside of domestic dwellings. In many respects, their presence has helped prevent numerous fires—much better to tackle the fire at the point of origin than to let it develop.

But there is a problem, and one born from a combination of health and safety policy, fire safety training, and litigation. The question we all need to ask ourselves is, do we expect persons who happen to be near the fire when it ignites to fight the fire with a portable fire extinguisher or a hose reel? If so, would those persons have been trained to use the equipment?

Are we asking too much of those persons, from a health and safety perspective, to stay and fight a fire rather than leave the building? What if they fail or get injured, or worse. Where does the employer stand legally?

Is it often the case that first aid firefighting equipment is supplied simply to meet national legislation, and it is there more for show than to form part of a fire strategy?

Despite this counterargument for first aid firefighting equipment, I do believe that it can play an invaluable part in a fire strategy, particularly where other parts of the strategy may not be robust enough. This therefore relies on an effective fire safety management regime that takes into account the questions raised here, together with any others that may be relevant to the building, its processes, and its occupancy.

Scoring the Factors

Once a number is chosen for each of the eight key strategic factors, they should be marked on the diagram (zero to five). Once all eight factors have been scored, straight lines should join the marks, and a pattern will emerge. This pattern can be quite revealing. See an example of a completed diagram in Figure 7.

Let us start with the area within the pattern. What does this tell us?

This gives an idea of the resources and costs associated with the provisions required by the strategy. For instance, if each element is scored as five, the pattern will form the whole outer core of the diagram, with the maximum area taken up. Consequently, the fire safety and protection provisions are likely to be costly to implement and extremely resource-hungry. Conversely, a shape with a much smaller footprint will be much more affordable. However, it must still be effective and answer all the issues from the earlier objectives setting.

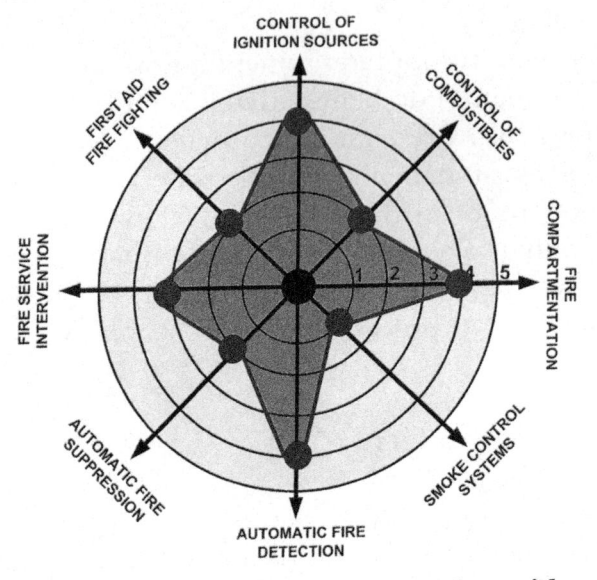

Figure 7: Use of fire strategy value grid

The way the pattern sits on the diagram also can tell something about the type of strategy. For a pattern predominantly in the upper quadrant, the strategy will place greater reliance on fire safety management, whilst a pattern towards the lower quadrant will place greater reliance on active fire protection. A shape on the left hand side indicates that the strategy will rely

STRATEGIC VISION

on suppression of a fire, whilst a shape to the right places greater reliance on containment and control of a fire and the products of combustion by structural means.

If the diagram has been prepared following a full prior evaluation, and by a team rather than a single person, then it is likely to be the precursor to the final written strategy. An item scoring highly in the diagram is likely to be an important feature of the final written strategy.

What those working on the diagram also realize is that the exercise promotes pure strategic thinking without getting into the detail.

CHAPTER 10.
USING YOUR RESOURCES

"That's been one of my mantras—focus and simplicity. Simple can be harder than complex: You have to work hard to get your thinking clean to make it simple. But it's worth it in the end because once you get there, you can move mountains."
—Steve Jobs, founder of Apple Computers

Smoke detection systems, heat detection systems, flame detection systems, multiple-sensor detecting systems, aspirating detection systems, video smoke detection systems, linear heat detection systems, open protocol systems, closed protocol systems, alarm sounder systems, public address systems, bell systems, fire telephone systems, manual smoke extract systems, forced extract systems, air pressurization systems, fire curtain systems, fire compartmentation structures and systems, fire and smoke dampers, fire doors, fire-rated glazing, fire sprinkler systems, wet riser and dropper systems, dry riser and dropper systems, firefighting hydrant systems, firefighting monitor systems, water deluge systems, gaseous fire extinguishing systems, firefighting foam systems, chemical fire extinguishing systems, inerting fire extinguishing systems, hose reel systems, portable fire extinguishers, fire service intervention, private firefighters, fire prevention policies, no smoking policies, flammable material choices, hot work controls...The list could go on!

The choice of resources available to implement a fire strategy allows for a range of options for any particular fire scenario. This, however, does not mean that a fire strategy can or should use a belts-and-braces approach.

The greatest military strategies optimised the use of limited resources to achieve success. Let's face it; any strategy will be successful if it has access to limitless resources. The same applies to business. Infinite funds, and access to anywhere, will inevitably allow the possibility of world domination. In fact, there really is no need for a thought-out strategy at all, as simply applying resources, without end, will eventually lead to success; whatever that may look like. To make the best use of limited resources, it really is all about focus and simplicity—thank you, Steve Jobs!

The following pages are provided to give some thoughts about how to determine what you *really* need to resource your fire strategy. The chapter will not describe all the resources available to fire strategists, as it is imagined that they already have a pretty good idea how the systems work and how they will perform in the event of a fire. If more information on system operation is required, a simple search on the Internet will quickly provide more information than you will ever need, as well as a variety of views. Try the growing number of blog sites on various issues to do with fire safety and protection.

USING YOUR RESOURCES

I would like to start with the issue of over-specification, over-engineering, or more specifically for the use of fire protection systems, "overprotection." In the days of total reliance on prescription, a fire strategy would be based on specific regulations and codes. This has key benefits in that the process is transparent and auditable, but the bad news is its inflexibility to modern building design and, more so, that such guidance is probably, and largely, risk-averse.

As highlighted earlier in this book, codes and rules are drafted by committees who, when required to compromise, will normally tend to default to the "safer" option. Consequently, there will be examples in many prescriptive codes, if followed religiously, that could lead to possible over-engineering. I gave an example in a previous chapter of the UK's sub-surface railway regulations. As a reaction to a major fire, everything was incorporated into this piece of legislation, leaving little room for movement and certainly a reduced ability to provide an optimum and cost-effective solution.

Now, with performance-based engineering, the fire strategist has the option of seeking to provide an optimum solution, minimizing the need for the requirement of more resources than is strictly necessary. There are, naturally, forces negating such optimised solutions, some coming from the increasingly imposing world of litigation. To some, over-engineering means more safety for the fire consultant. Perhaps we need to be somewhere between these poles.

The concept of over-engineering or overprotection can be illustrated by the curve in Figure 8, the marginal value of fire precautions. Note that the diagram plots the marginal value of adding fire precautions against cost and breaks the curve into four sectors.

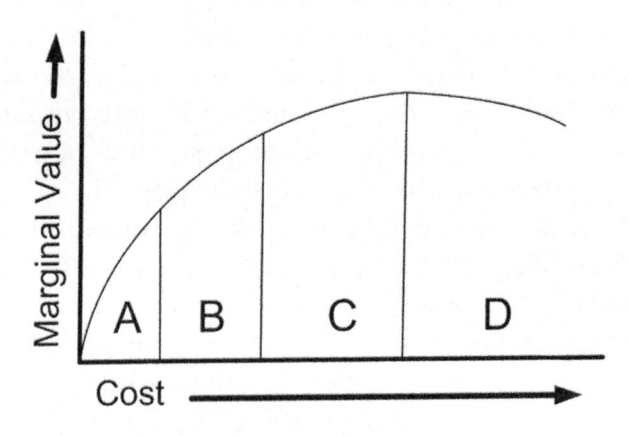

Figure 8: Marginal value of fire precautions

Sector A is the point where, for a little outlay, basic fire precautions can achieve credible results to the fire strategy. Anything from fire prevention housekeeping procedures to smoking bans (now not such an obstacle) to good all-round fire safety management could be applied at this point. As this would suggest, effective fire safety management can potentially improve the overall fire safety dynamics of a fire strategy. I have seen very good fire strategies that rely entirely on the actions of people rather than the actions of systems.

Sector B is probably where good fire safety management is supplemented by more sophisticated management processes and policies combined with the introduction of fire protection systems. Whatever has been specified is still valuable and contributes to an

USING YOUR RESOURCES

effective fire strategy. This is probably the point where fire detection and alarm systems are introduced.

Sector C is where every fire strategy wants to be—sufficient combinations of fire safety management and fire protection systems to meet all the objectives, with every system making a positive contribution.

Sector D is where I have seen some fire strategies end up. Defining the tipping point between C and D is quite difficult to do, but when you start to ask a few of the "three why" questions as described earlier in this book, you may decide that we have reached D simply because some of the systems do not seem to add any function to the overall fire strategy.

> **How would you determine if the fire strategy solution incorporates over-engineering?**

Over-engineering or overprotection may not always be a result of risk-averseness or even incompetence. Sometimes there may be a corporate policy for over and above fire protection, based on previous experience of fires—"I am not chancing that again" There may be situations where there is not complete confidence in part of the provisions, so some doubling up is deemed appropriate, and there is sometimes a general reluctance to take any form of risk.

There is also the very real possibility that stakeholders in the strategy, whether part of the team or outside the team, make additional demands of the strategy than strictly necessary. In such cases, the fire strategist will need to make a clear argument with technical back-up. A well-reasoned and thorough response is very hard to fault.

Once the performance criteria of the strategy have been developed, there will be a need to allocate resources to meet those criteria. Every type of fire protection system will have its strengths and weaknesses. Some systems will be better than others under specific conditions and in a certain set of circumstances. There are three key variables each system can be measured by: technical performance, logistics, and economic considerations.

Technical performance: Technical performance is the ability for a type of system or process to reliably perform its function. Features of the building could restrict choice of one type of system or process and could favour others.

Logistics: This is where certain fire provisions and systems may be desirable, but the nature of the building, its process, and its people means that the logistics are likely to discount them as viable options. Features of the building that physically restrict some aspects of the strategy are a good example. The ease of maintenance is an important issue that must be taken into account, possibly at the strategic rather than tactical level, even though siting of devices is not normally part of a fire strategy. If the logistics are such that any part of a system cannot be maintained, then an alternative solution needs to be sought.

Economic considerations: Any fire strategy must be commercially viable, and this will include the cost of the resources used to meet the needs of the strategy. Whole lifetime costs of a solution need to be established.

Understanding and highlighting the constraints on the strategy should be a focus at an early stage of its preparation. Consideration will need to be given to the maintainability of the strategy over the long term. An inappropriate or overly complicated strategy is likely to be misinterpreted, underused, or "binned." This in itself should be regarded as a constraint. Note that decisions with regard to constraints should be made in consultation with all interested parties.

In order to overcome constraints, the fire strategist needs to think in a purely objective way. One of the problems of being objective is that we all have ideas as to what works best and what, based on our experience, works. Combine this with the constraints identified earlier and we all, unintentionally, limit the choice of resource based on our own ideas as to what can work, and what will not work.

Experience, with all its benefits, can also lead us to rely on tried and trusted solutions rather than one that may be most appropriate for the circumstances we are faced with. To help introduce objectivity into choice, I developed a tool to assist in determining the most appropriate option for a given set of conditions. The idea is that, when faced with a choice of strategy options, we need to put subjectivity to one side and score each of the possible ways forward. There could be two, three, or more options, and each option has certain strengths and weaknesses when required to fulfil a part of the fire strategy for that specific building.

I determined that the applicability of each option should be based on the three factors given above.

I also realised that, for different projects, the relative importance of each of the factors will change. For some projects, technical performance may far outweigh logistical and economic factors. In other cases, economic considerations will be deemed to be most important, particularly if each of the choices could change the budgets considerably.

The table is shown in Figure 9. The idea is that each option is scored on each of the three factors. You then multiply the three scores together for each option and arrive at a number for each option. The higher the number, the most appropriate the option is. I did think about using addition rather than multiplication, but the results will be less obvious.

CRITERIA	OPTION A	OPTION B	OPTION C
Performance (Score from x)	N1 out of x	N2 out of x	N3 out of x
	Multiply by ...	Multiply by ...	Multiply by ...
Logistics (Score from y)	N4 out of y	N5 out of y	N6 out of y
	Multiply by ...	Multiply by ...	Multiply by ...
Economics (Score from z)	N7 out of z	N8 out of z	N9 out of z
TOTAL SCORE	=	=	=

Figure 9: Quantified assessment of options

Note that each of the factors incorporates a separate scoring system such that the one that is most important will be scored from a higher number than the others. The choice of number is arbitrary but, for instance, if logistics is the most important, a maximum score of 10 could be offered whilst for the least important, economic, a maximum score of 5 could be used. In this way, the importance of the factor will skew the result to take into account the importance of certain factors.

This tool is by no means meant to be an absolute method for ensuring that the right option is chosen. It is there to prompt discussion and to enable the thinking process.

Another way to look at the use and necessity of resources is to consider whether those resources are strictly necessary and contribute to the fire strategy. This makes use of a tool I developed way back in the 1980s— the one I formulated as part of a fire insurance working group in Paris (see Preface). It is designed to be simple in its approach and is, in effect, a flow chart that starts with the detection of a fire and ends with one of two fundamental objectives: to get the people out and to put the fire out. The point of this exercise is to determine, as part of a sequence of events, what key resources are required. The diagram is shown in Figure 10.

The starting point is a fire that is detected. Note that we would assume that the method of detection is an automatic fire detection system, but this need not be the case. If the only method of detection is via the human eye and nose, then this will still work. In fact, if the strategy does not rely on systems to detect fire, then there is an immediate technical resource saving, although the fire safety management strategy will need to cover this.

FIRE STRATEGIES - STRATEGIC THINKING

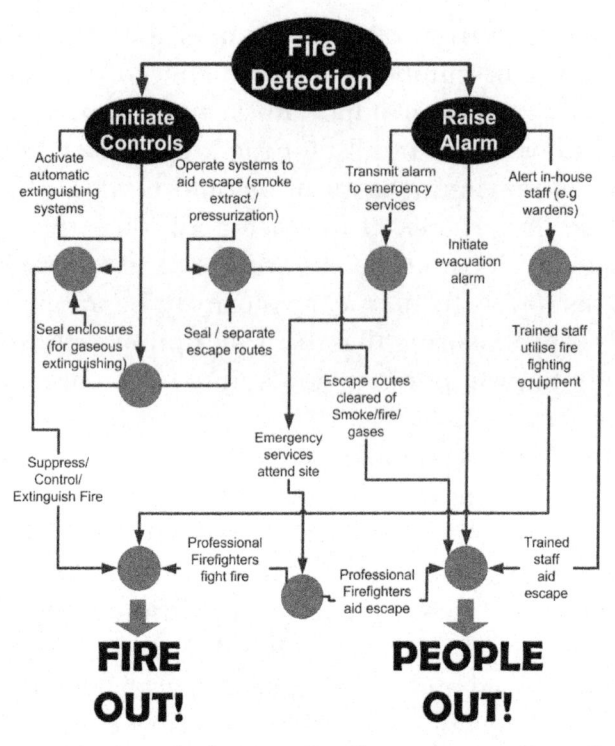

Figure 10: Components of a fire strategy flow chart

Fire detection is there to provide for two main functions; to initiate warning systems and to initiate control systems. If we took a simplistic view, we could judge that the prime purpose of warning systems is to get the people out and the main purpose of control systems is to control the fire. This is true, but only to a point. The whole purpose of this diagram is to show how warning and control systems can work together to collectively achieve the end objectives.

Let us move down to the next level. Three forms of control are shown as the activation of fire extinguishing

USING YOUR RESOURCES

systems, the operation of systems designed to prevent smoke and fire spread (this will include fire doors and fire and smoke dampers), and the operation of systems designed to aid escape (smoke extract systems, space pressurization systems, and so on). It can already be noted that two of the three systems are designed for life safety. The systems designed to prevent smoke spread may also play a key part in assisting with fire extinguishing (such as maintaining a tight enclosure for gaseous systems) as well as ensuring the safety of escape routes and assisting smoke control systems by forming tight enclosures or operating features such as fire curtains.

Moving to the warning systems, once again we have three possible outputs. The main output will be initiating evacuation alarms (this may also include alert sequences), alerting in-house staff (possibly via coded warnings or other forms of communication), and alerting the emergency services via remote transmission systems. Once again, two of these three outputs will assist with the "fire out" objective, whether extinguishment can be achieved or supported by the emergency services or by in-house staff. Note that the emergency services and in-house staff may also play a pivotal role in aiding the evacuation.

The diagram possibly simplifies the arrangement of systems, particularly for complex buildings and environments, but it does help concentrate the mind on picturing the sequence of events and how systems and operations play their part. Sometimes, if the strategy gets stuck and the overall ideas get lost, a simple pictorial arrangement may help in refocusing the mind of the fire strategist.

FIRE STRATEGIES - STRATEGIC THINKING

The purpose of this chapter is to provide thought when it comes to the use of resources. Fire safety and protection is blessed with so many different types of system and methods to achieve specific objectives. It is up to fire strategists to focus their attention on what the strategy is aiming to do and choose the best and most appropriate systems and processes to meet those aims.

It can be very tempting to over-engineer, particularly as the pressures of providing a strategy to save life, property, business, and the environment can be complex. The temptation to provide more than is really necessary is understandable, particularly in this increasing world of litigation. A strategy that overprotects is not just a matter of additional cost; it may also overcomplicate the fire safety provisions and lead to a strategy that will inevitably decay to a state where it is no longer viable. A tough call for the fire strategist? Maybe, but a steady focus on the key objectives should lead to a strategy that optimises resources rather than complicates them.

We considered another trait amongst many in the fire safety provision, and that is reliance on our favourite system arrangements. I do it, and many others I know do it. We get used to arrangements that work and thus tend to go on trusting the arrangements time and time again. Subjectivity is not always a bad thing, but it can stifle new ideas. Bringing a degree of objectivity into the choice of options can open doors to new concepts.

CHAPTER 11.
HEADING TO VICTORY

"Accept the challenges so that you can feel the exhilaration of victory."
—George S. Patton, Second World War US General

Victory is when you meet your objectives. It is the point when you know you have completed your assignment and achieved what you had set out to do. But what does victory look like for a fire strategist, and how should it look to the other stakeholders?

I remember being involved as the client in a tendering process for the preparation of a fire strategy for a complex building. I also remember that the returned tenders varied so widely that you were left wondering if any of the bidding organisations knew what they were to supply. What should a fire strategy document look or feel like?

I suppose this starts with what the expectation is from each of the involved stakeholders. It questions how much effort is required to prepare a fire strategy and how deep the fire strategists will go in providing a strategy worthy of the building. In summary, what exactly will the fire strategy cover and what level of analysis will be undertaken?

It is not unusual that those competent to prepare fire strategies will have an in-house style and a format that they have used in a variety of cases. This is not really a problem, but it would be useful for those bidding to write a fire strategy to explain in some detail their approach to the problem, what exactly they will do, and what the final delivery will be. As highlighted in this book, a fire strategy may be an opportunity for a more comprehensive analysis than one that delivers what those issuing the commission *believe* they need. In many cases, they are not the experts, and it is thus appropriate and ethical for fire strategists to explain their take on the situation, preferably as part of the tender documentation or at least in the early stages of the project—an opportunity often missed or wasted.

> *Do you believe all stakeholders understand what fire strategy preparation involves?*

The proposed style could be in the form of a contents page with a brief description of what each of the contents will cover. It is also likely that the style and arrangement of the document will differ whether a prescriptive or performance-based approach is utilised. Suitable formats could follow national fire regulations or codes.

One of the main reasons why I decided to write a code for fire strategies was to introduce consistency in approach. After seeing too many formats and levels of detail, I thought that a typical format that could be applied internationally would create a much more even platform, particularly for a tendering process. It all starts with the premise that a fire strategy document should simply be clear and concise and that the conclusions reached should be clear to all stakeholders even if they do not understand the accompanying detail.

The strategy should ideally be a single overview of all relevant fire precautions specific to the building or infrastructure subject to the strategy. It should be pitched at the right level—neither vague nor overly detailed into the tactics to be used. It has to be flexible to allow for alternative approaches, but not such that anything could effectively be used, a result of ambiguity.

It is not meant to specify detailed designs or arrangements but is there to give a sufficient framework for more detailed assessments and designs. In this way, it is likely to continue to be relevant throughout minor changes that may affect its shelf life. Examples of inappropriate wordings are:

The fire detection system should be installed as appropriate. (What does this mean exactly?)

All fire separations should be constructed using two layers of "Manufacturer" board on a three-inch thick wooden frame. (This is a specification and not a strategy)

A fire strategy should cover, for example, the layout and protection of escape routes, the extent of detection, the areas that require fire suppression, the level of fire safety management, and the type and extent of warning systems. In fact, a fire strategy should include consideration of all aspects related to fire safety and protection.

There are sometimes cases where features do not seem to have been covered. Could this be the result of the fire strategist deciding not to include some aspects in the strategy? Did he or she plainly not include them for other reasons, such as those related to memory?

How do you know, absolutely, that because, say, smoke control is not covered in the strategy, it has been deemed irrelevant, or impractical? Could it be that smoke control is not included because another solution is included that covers the same objective, or that smoke control is not required for other reasons? I find that the "cut and paste" approach is a good way for omissions to be introduced—by accident.

In order to be sure that every aspect that may or may not be relevant has been considered, it may be a good idea to break down a full strategy into a series of sub-strategies. In this way, it is possible to identify if and where a single element of a fire strategy is relevant or not, as the case may be. I used this sub-strategy approach in BS PAS 911, and I will include it again in this chapter to assist with strategy formulation. Note that the PAS incorporates a table for each of the sub-strategies, with a tick box, so that each element of each sub-strategy is separately identifiable and is either included in the strategy or is deemed to be not applicable. The possibility that an element is forgotten about by the fire strategist should thus be minimized. It also gives other stakeholders the opportunity to ask pertinent questions of the fire strategist.

The sub-strategy approach is there mainly to find if there are holes, and it is not proposed that the layout and format of the strategy document follow this method. Then again, why not if you do not have a format in mind?

Before we identify the sub-objectives I firmly believe that every fire strategy should start by including a statement—an introduction providing an overview and basis for the strategy and identifying the key objectives for the strategy. The fire strategy statement could form part or the entire introduction to the strategy. The statement will, in fact, be the historical perspective; thus, those reviewing the strategy at a later date will be able to understand the background and reasons for how the strategy was derived. A good starting point would be to identify the stakeholders involved in the exercise. This has a twofold benefit in that it provides for an audit trail for decisions made and it assures the reader that not just one perspective has been taken into account.

The statement should not just include the objectives set by the strategy but should also give additional information as to the objectives setting. For instance, the statement could go much further than to state only that the strategy covers life safety. It could go a little further into the occupancy profiles covered. A statement suggesting that it is for

> **From your experience, do fire strategies clearly state what their objectives and scope are?**

property protection could be embellished by stating key features of the property profile being protected—not just the building but possibly key contents of the building, and so on.

The statement is also a good opportunity to cover key assumptions made and the criteria behind the decisions made. Areas that could not be properly assessed could be highlighted at this stage in the strategy report. The statement, in fact, sets the stage for the remainder of the document.

The fire safety (management) strategy will be the first of the sub-strategies and will initially cover how fire safety is to be managed in the building. For existing buildings, this could include details of how the current regime manages fire safety, highlighting those deemed responsible, how they sit in the overall structure, and the levels of responsibility and authority given. The strategy could then cover the methods used, as applicable, for fire prevention, such as the use of materials, controls on ignition sources, and general housekeeping arrangements, including the assessment of fire risks.

It should also cover how the strategy is to be maintained. This will include the methods used to ensure continued compliance with the strategy, including procedures to ensure maintenance and continued efficacy of fire protection systems—key where the strategy relies upon such systems. Issues such as training should also be considered at this point.

What is listed here is just a taster of what proper fire safety management may include. There are a number of national standards and guidance documents that can provide additional information. Furthermore, as fire safety management can ensure the success of any type of fire strategy, what is said in this sub-strategy will set the basis for every aspect of the overall fire strategy.

A key sub-strategy that will be fundamental for every form of life safety strategy, and even where the safety of occupants is not a key driver, is the evacuation strategy. This will cover the fundamentals of the means of escape, incorporating the identification of all escape routes together with the methods used for their protection, illumination, and signage. It will identify the criteria for both horizontal and vertical escape. Where escape directly to a place of safety is not possible, the evacuation strategy should identify alternative places of relative safety including refuges and how these should be used. Note that, as highlighted earlier, refuges will need to link directly with the management of fire safety, as without these arrangements covered, the strategy is unlikely to succeed. Naturally, the escape of mobility-impaired persons will also need to be covered unless it is specifically excluded from the strategy by stakeholder agreement.

The evacuation sub-strategy should consider how persons are alerted to the fire and how the evacuation process is controlled to ensure optimum escape strategies. This will cover the type of warning systems and how they are to be utilised. If coded or staged evacuation systems are specified, then the strategy should state how they are used and describe the sequence of events. Where appropriate, the way that pre-movement time is minimised could be included here. Note that there will once again be a strong link with the management systems employed.

Fire and smoke control is, in many cases, fundamental to the success of other sub-strategies and accordingly should be considered as its own sub-strategy. Every strategy should be able to state the methods adopted to control the external spread of fire via both walls and roofs. Similarly, the methods used to control the internal spread of fire should be identified. This should look at the subject from both the spread via linings (linking once again to some of the fire safety management criteria) and via the structure, from stability of the building itself and the impact of connected buildings through to the arrangement of internal fire compartments.

Ideally, the fire strategy will utilise drawings showing both horizontal and vertical fire separations, including the basis used for determining the compartmentation criteria. Note also that where there is a possibility of concealed spaces where fire and smoke could conceivably travel, this should be covered within the strategy.

Fire and smoke control may also rely on active systems, and these systems should be explicitly stated together with any performance criteria and purpose. If the systems rely on complementary measures, such as fire curtains, then this should be stated. Note that where manual means are required to extract smoke, this should be incorporated into the fire safety management sub-strategy.

The firefighting sub-strategy will encompass two key elements: "first aid" firefighting and "professional" firefighting. Although first aid firefighting systems are sometimes an afterthought within a comprehensive strategy, their function should be incorporated. Note that first aid firefighting training may be an ingredient in the management systems.

When it comes to professional firefighting, whether it relies on a private, community, or national emergency response service, the strategy will need to not just cover the practical issues for fire service attendance but also the infrastructure required to properly fight fires. The attendance criteria need to be clearly stated. Practical issues such as vehicular access, escort arrangements, and access to equipment will need to be considered.

The fire strategist will also need to identify if the firefighting systems are adequate to meet with the expected demand functions both based on the physical requirements for firefighting and to provide for the objectives set. It is not unusual for major upgrades of firefighting water installations to be deemed necessary simply to meet the demand required. Where environmental considerations are deemed appropriate, issues such as water or foam run-off from fighting a fire should be addressed.

The final sub-strategy is that covering active fire protection systems: the fire protection sub-strategy. This will cover detection and suppression systems and any other active systems that support the overall fire strategy. Let us start with the fire detection system. Ideally, the fire strategy will state the prime purpose of the detection system, the areas of coverage, and the type(s) of detection deemed necessary. It should then

FIRE STRATEGIES - STRATEGIC THINKING

describe the layout and configuration requirements for control and indicating equipment and how the systems interact with other systems. Although it is not expected that a fire strategy will detail cause-and-effect logic, it should provide an overview of how detection in one area affects outputs in that area or in other areas. It should, for instance, state that a detector operating in Area 1 operates evacuation sounders in Areas 1, 2, and 3 and an alert message in Areas 4, 5, and 6. It should determine the operability of systems used to maintain fire compartmentation. In fact, it should include enough information to allow a specialist fire systems engineer to design cause-and-effect logic.

Then there are fire suppression systems that may include anything from water sprinkler systems to deluge systems to gaseous extinguishing systems. The strategy should identify the type(s) of system required, their location, and key operational criteria. Although the design of the system will not form part of the strategy, the fire strategist should ensure that the infrastructure is capable of allowing for the installation, which will include consideration of the location of water tanks and gas cylinders and, where appropriate, the potential air tightness of enclosures (although the fire strategist will not be expected to undertake room integrity testing). Although detection and suppression systems have been identified, this sub-strategy applies to *any* other active systems required to assist with the fire strategy.

As all active fire protection systems require maintenance and need to regularly reassessed for efficacy, this should be highlighted within the strategy document.

Note that each sub-strategy may incorporate elements of other sub-strategies so there may be, in fact there *should* be, some repetition. This is a good thing in that it helps ensure that no elements of the strategy have been overlooked and that each of the sub-strategies is linked. And that is largely the point of this book—that all the elements of a strategy contribute towards a single goal, a single plan to meet the objectives set. The elements work together to provide a consistent and complementary set of measures.

The benefit of using the sub-strategy approach is that it allows for revisions of one aspect of the overall strategy without reviewing all parts. For instance, changes to occupancy numbers may primarily affect the evacuation strategy. Therefore, this sub-strategy could be reviewed without affecting all other parts. Note that where there are common parts with other sub-strategies, these also will need to be reassessed.

Before the strategy is released to the client, it is highly useful for the fire strategists involved to take time to review their work. This may involve peer review. Colleagues not involved in the day-to-day aspects of fire strategy preparation for the project could give a good, objective overview and identify ambiguity and areas that may have been missed. In some cases, peers could be others working in the profession but not in the same organization. This may become more common as fire consultancy companies develop partnership arrangements. The strategy should also be signed off by someone senior in the organization.

Once the client receives the fire strategy, they should be reminded that the strategy should be reviewed on a periodic basis and following any changes that are likely to have an impact on it. It is recommended that those with responsibility for the strategy undertake an in-house review annually. It is also recommended that a more formal review be undertaken at least every five years by persons competent to prepare strategies.

Extreme events

A topic that I have not touched on, but that may be referred to in a fire strategy, is that of extreme events. What comes to mind when we think about extreme events is possibly something similar to the 9/11 atrocity in New York in 2001. How could any fire strategy cope with such an extreme set of circumstances? Would it even be conceivable that a fire strategy could and should cater for such circumstances, even if a scenario of a small plane colliding with one of the towers may have actually been analysed?

Let us not forget that, when we do look at scenarios, even two simultaneous fires starting in different parts of a building could be regarded as an extreme event. Legislation and guidance are often based on the premise of a single fire event at any one time. Could you

> *Do you believe that there are some extreme events that should be incorporated into fire strategies?*

imagine the complexity of fire precautions necessary when assuming two or more simultaneous fire events as a minimum?

The fire strategy report should therefore highlight somewhere within it that it does not cover extreme events. That is, unless the client has highlighted that certain events should be taken into account and would be happy with the cost of additional provisions that would probably be necessary.

But we must never forget that, in some cases, extreme events will occur, and it is less to do with holes in the fire strategy or with any related strategy, such as a security plan, than how we handle the situation. A crisis management plan is designed to ensure that, if and when a crisis situation occurs, an organisation has the relevant systems in place to respond quickly and efficiently, with the objective to limit the impact and duration of the crisis, and to prevent secondary crises occurring from inappropriate actions undertaken whilst handling the original situation. The plan is to provide additional degrees of resilience to an organization, and this will be over and above the fire strategy, security strategy, and so on.

From 2007 to 2009, my company, Kingfell, looked into the relationship between fire safety and protection and crisis management. In effect, the crisis management plan will pick up where the fire strategy leaves off and will be a set of procedures covering preparation, response, and recovery. Crisis management is not just about events such as major fires or other catastrophes. A crisis could be an adverse set of economic data. It could be corporate defamation, or it could be an unusual set of environmental circumstances working together to cause a major problem to the organisation.

In fact, a fire could be the result of a set of events, from intentional or accidental actions to the impact of adverse weather conditions. Sometimes identifying potential threats to an organisation, wherever they come from, can lead to plans to mitigate them and thus, in turn, to reduce, in this instance, the fire threat.

This could well be a topic for a future book, as I believe that all fire strategists have something very much in common with others involved in managing crises. Perhaps in thinking this way, fire, security, and crisis management experts may come together and form overall safety, protection, and emergency management strategies. Now that *really* would be strategic thinking!

I do hope that you found this book of use. The experience of writing it has helped me change the way I see the people within this profession. There is so much experience and expertise in the fire sector, so many interesting people with interesting views. Fire engineering, or fire safety engineering, really is a beautiful subject!

REFERENCES

The following British Standards are referred to in this book and are available from http://shop.bsigroup.com:

British Standard Specification: PAS 911: 2007: Fire strategies—guidance and framework for their formulation.

British Standard (BS) 5839-1:2002+A2:2008: Fire detection and fire alarm systems for buildings. Code of practice for system design, installation, commissioning and maintenance.

British Standard (BS) 7974:2001: Application of fire safety engineering principles to the design of buildings. Code of practice.

British Standard (BS) 9999: 2008: Code of practice for fire safety in the design, management and use of buildings.

ABOUT PAUL BRYANT

Paul is a British chartered fire engineer and chartered electrical engineer. He has a first degree in electrical engineering and a master's degree in business administration.

He started his career in the early 1980s with a UK-based fire insurance organization known as the Fire Offices' Committee. He was involved with the approval of fire detection systems and represented British fire insurers in Europe and at an international level. He then moved on to the UK's Loss Prevention Council and was involved with the writing of insurance fire standards.

In the early nineties, Paul joined London Underground Limited as head of fire engineering and took on the audit role of a major fire compliance project, upgrading the fire detection, suppression, and compartmentation systems in 115 subsurface railway stations.

He formed Kingfell in 1995 and grew it into a major fire consultancy and contracting business before reshaping it in 2011 to concentrate on complex building fire strategies.

Paul has been involved in fire standards making from the early days in his career. He has drafted a number of standards for the fire insurance sector as well as British and European Standards. He was one of the youngest British Standards' Technical Committee chairmen at the age of twenty-five.

Paul is a Freeman of the City of London and is a member of the Worshipful Company of Firefighters

Paul is now developing his business in Europe, the Middle East and the USA.

Printed in Great Britain
by Amazon